T0390914

CAMBRIDGE TRACTS IN MATHEMATICS

General Editors

147 Floer homology groups in Yang–Mills theory

Floer homology groups in Yang–Mills theory

S. K. Donaldson
Imperial College, London

with the assistance of M. Furuta and D. Kotschick

CAMBRIDGE UNIVERSITY PRESS
Cambridge, New York, Melbourne, Madrid, Cape Town, Singapore,
São Paulo, Delhi, Dubai, Tokyo, Mexico City

Cambridge University Press
The Edinburgh Building, Cambridge CB2 8RU, UK

Published in the United States of America by Cambridge University Press, New York

www.cambridge.org
Information on this title: www.cambridge.org/9780521808033

First published 2002
Reprinted 2003

A catalogue record for this publication is available from the British Library

Library of Congress Cataloguing in Publication Data

Donaldson, S. K.
Floer homology groups in Yang–Mills theory / S. K. Donaldson with
the assistance of M. Furuta and D. Kotschick
p. cm. – (Cambridge tracts in mathematics ; 147)
Includes bibliographical references and index.
ISBN 0 521 80803 0
1. Yang–Mills theory. 2. Differential geometry.
I. Furuta, M. II. Kotschick, D. III. Title. IV. Series.
QC174.52.Y37 D66 2001
530.14'35–dc21 2001035888

ISBN 978-0-521-80803-3 Hardback

Contents

1	**Introduction**	*page* **1**
2	**Basic material**	**7**
2.1	Yang–Mills theory over compact manifolds	7
2.2	The case of a compact 4-manifold	9
2.3	Technical results	10
2.4	Manifolds with tubular ends	13
2.5	Yang–Mills theory and 3-manifolds	14
	2.5.1 Initial discussion	14
	2.5.2 The Chern–Simons functional	16
	2.5.3 The instanton equation	20
	2.5.4 Linear operators	23
2.6	Appendix A: local models	27
2.7	Appendix B: pseudo-holomorphic maps	30
2.8	Appendix C: relations with mechanics	33
3	**Linear analysis**	**40**
3.1	Separation of variables	40
	3.1.1 Sobolev spaces on tubes	45
3.2	The index	47
	3.2.1 Remarks on other operators	51
3.3	The addition property	53
	3.3.1 Weighted spaces	58
	3.3.2 Floer's grading function; relation with the Atiyah, Patodi, Singer theory	64
	3.3.3 Refinement of weighted theory	68
3.4	L^p theory	70

4	**Gauge theory and tubular ends**	**76**
4.1	Exponential decay	77
4.2	Moduli theory	82
4.3	Moduli theory and weighted spaces	87
4.4	Gluing instantons	91
	4.4.1 Gluing in the reducible case	100
4.5	Appendix A: further analytical results	103
	4.5.1 Convergence in the general case	103
	4.5.2 Gluing in the Morse–Bott case	108
5	**The Floer homology groups**	**113**
5.1	Compactness properties	113
5.2	Floer's instanton homology groups	122
5.3	Independence of metric	123
5.4	Orientations	130
5.5	Deforming the equations	134
	5.5.1 Transversality arguments	139
5.6	$U(2)$ and $SO(3)$ connections	145
6	**Floer homology and 4-manifold invariants**	**151**
6.1	The conceptual picture	151
6.2	The straightforward case	158
6.3	Review of invariants for closed 4-manifolds	161
6.4	Invariants for manifolds with boundary and $b^+ > 1$	165
7	**Reducible connections and cup products**	**168**
7.1	The maps D_1, D_2	168
7.2	Manifolds with $b^+ = 0, 1$	169
	7.2.1 The case $b^+ = 1$	171
	7.2.2 The case $b^+ = 0$	174
7.3	The cup product	176
	7.3.1 Algebro-topological interpretation	176
	7.3.2 An alternative description	179
	7.3.3 The reducible connection	183
	7.3.4 Equivariant theory	188
	7.3.5 Limitations of existing theory	196
7.4	Connected sums	201
	7.4.1 Surgery and instanton invariants	201
	7.4.2 The $\mathrm{Hom}_{\mathcal{F}}$-complex and connected sums	206
8	**Further directions**	**213**
8.1	Floer homology for other 3-manifolds	213

8.2 The blow-up formula 219

Bibliography 231
Index 235

1
Introduction

In 1985 Andreas Floer discovered new topological invariants of certain 3-manifolds, the 'Floer homology groups'. This book originated from a series of seminars on this subject held in Oxford in 1988, the manuscript for the book being written sporadically over the intervening 12 years. The original plan of the project has been modified over time, but the basic aims have remained largely the same: these are, first, to give a thorough exposition of Floer's original work, and, second, to develop some further aspects of the theory which have not appeared in detail in the literature before. The author can only apologise for the long delay in completing this project.

Floer's original motivation for introducing his groups – beyond the intrinsic interest and beauty of the construction – seems to have been largely as a source of new invariants in 3-manifold theory, refining the Casson invariant which had been discovered shortly before. It was soon realised however that Floer's conception fitted in perfectly with the 'instanton invariants' of 4-dimensional manifolds, which date from much the same period. Roughly speaking, the Floer groups are the data required to extend this theory from closed 4-manifolds to manifolds with boundary. From another point of view the Floer groups appear, formally, as the homology groups in the 'middle dimension' of an infinite-dimensional space (the space of connections modulo equivalence) associated to a 3-manifold. This picture is obtained by carrying certain aspects of the Morse theory description of the homology of a finite-dimensional manifold over to infinite dimensions. All of this is closely related to ideas from quantum field theory – indeed, one of Floer's starting points was the renowned paper of Witten, [49], which *inter alia* forged a link between quantum mechanics and Morse theory

– and the connection with mathematical physics permeates the whole subject.

The formal properties of the Floer groups, and their relation with invariants in four dimensions, fit into a general conceptual framework of 'topological quantum field theories' which was propounded in the late 1980s by Segal, Atiyah, Witten and others. We recall from [2] that a topological field theory, in $d+1$ dimensions, consists of two functors on manifolds. The first assigns to each closed, oriented, d-manifold Y a vector space $\mathcal{H}(Y)$ (over, say, the complex numbers). The second assigns to each compact, oriented $(d+1)$-dimensional manifold X with boundary Y a vector

$$Z(X) \in \mathcal{H}(Y).$$

These are required to satisfy three axioms:

(1) The vector space assigned to a disjoint union $Y_1 \cup Y_2$ is the tensor product

$$\mathcal{H}(Y_1 \cup Y_2) = \mathcal{H}(Y_1) \otimes \mathcal{H}(Y_2).$$

(2) $\mathcal{H}(\overline{Y}) = \mathcal{H}(Y)^*$, where $\overline{Y}$ is Y with the reversed orientation.

(3) Suppose X is a $(d+1)$-manifold with boundary (which may be disconnected), and that X contains Y and $\overline{Y}$ as two of its boundary components. Let $X^\sharp$ be the oriented manifold obtained from X by identifying these two boundary components. Then we require that

$$Z(X^\sharp) = c(Z(X)),$$

where the contraction $c : \mathcal{H}(\partial X) \to \mathcal{H}(\partial X^\sharp)$ is induced from the dual pairing $\mathcal{H}(Y) \otimes \mathcal{H}(\overline{Y}) \to \mathbf{C}$ and the decomposition

$$\mathcal{H}(\partial X) = \mathcal{H}(Y) \otimes \mathcal{H}(\overline{Y}) \otimes \mathcal{H}(\partial X^\sharp).$$

These axioms have some simple consequences. First, Axiom 1 implies that if $Y = \emptyset$ is the empty d-manifold then $\mathcal{H}(\emptyset)$ is canonically isomorphic to $\mathbf{C}$. Thus if X is a *closed* $(d+1)$-manifold the vector $Z(X)$ is a numerical invariant of X. Second, suppose that a $(d+1)$-manifold U is a cobordism from Y_1 to Y_2, so the oriented boundary of U is a disjoint union $\overline{Y}_1 \cup Y_2$. Then, by Axioms 1 and 2, $Z(U)$ is an element of $\mathcal{H}(Y_1)^* \otimes \mathcal{H}(Y_2)$ and hence gives a linear map

$$\zeta_U : \mathcal{H}(Y_1) \to \mathcal{H}(Y_2).$$

If V is a cobordism from Y_2 to a third manifold Y_3 then Axiom 3 states that

$$\zeta_{V \circ U} = \zeta_V \circ \zeta_U : \mathcal{H}(Y_1) \to \mathcal{H}(Y_3),$$

where $V \circ U$ is the obvious composite cobordism. So we obtain a functor from the category of d-manifolds, with morphisms defined by cobordisms, to the category of vector spaces and linear maps.

The original motivation which led Segal and others to develop this kind of axiomatic picture was to abstract in a tidy mathematical form the basic structure of quantum field theories (more precisely, of conformal field theory on Riemann surfaces). The theories which are usually considered in physics differ from the set-up considered above in that they operate on manifolds with some additional differential-geometric structure, for example a Riemannian metric or a conformal structure. It is precisely the absence of these geometric structures in our set-up which leads to the designation *topological* quantum field theories, and which means that we obtain topological (or, more precisely, differential-topological) invariants of manifolds. In a typical physical set-up the corresponding space $\mathcal{H}(Y)$ would be an infinite-dimensional Hilbert space defined, at least schematically, by associating to Y a space of 'fields' $\mathcal{C}(Y)$ (an element of $\mathcal{C}(Y)$ might be a tensor field over Y), and then letting $\mathcal{H}(Y)$ be a space of L^2 *functions* on $\mathcal{C}(Y)$. The vector $Z(X)$ is obtained by functional integration over a space of fields on X, with given boundary value on Y.

The Yang–Mills invariants, and Floer groups, fit into this general scheme, with $d = 3$. In outline, for a 3-manifold Y, we take the Floer groups (with complex co-efficients say)

$$\mathcal{H}(Y) = HF_*(Y).$$

For a closed 4-manifold X the Yang–Mills instantons define a numerical invariant $Z(X)$, and for a 4-manifold with boundary we obtain invariants with values in the Floer homology of the boundary. Actually, as we shall see, the simple axioms above need to be modified slightly to apply to the Yang–Mills set-up and the theory has a number of special features. For example, the invariants of a closed 4-manifold are not in general just numbers but functions on the homology of the manifold – so we might regard the functor as being defined on a category of 4-manifolds containing preferred homology classes. Nevertheless these axioms capture the essence of the matter. Contrasting with the physical set-up outlined above, we can say very roughly that in place of the infinite-dimensional space $\mathcal{C}(Y)$ of *all* fields (i.e. connections) we restrict in this topological theory to the *finite* set of flat connections (modulo equivalence) over Y, and we restrict to 'instanton' connections over 4-manifolds, so that in place of the functional integration over connections we now have merely to *count* the instantons with given flat boundary values. To make rigorous sense of this, a key step is to add half-infinite tubes to our 4-manifolds, so that we have a picture in which the boundary is 'at infinity'.

An important goal then of this book is to develop this picture, of the Floer groups as part of a topological field theory, in detail. It is important to emphasise at the outset that, even after all this time, we are not able to complete this task. On the one hand there are, as we shall see, rather fundamental technical reasons why one cannot expect to have this simple picture without imposing some restrictions on the manifolds which are considered. On the other hand, even within the confines of the theory that one might reasonably hope for, there are crucial technical difficulties, arising from the non-compactness of instanton moduli spaces which – despite much labour by many mathematicians – have not yet been fully overcome. Failing, therefore, a definitive treatment we round off the book, in Chapter 8, by seeking to explain the problems that remain, and further developments one may expect in the future. We shall see that – far from being dull, technical matters – these difficulties lead to striking and unexpected formulae involving classical special functions.

Throughout the early 1990s an important motivation for the development of Floer theory was the hope that this might lead to new calculations of 4-manifold invariants, via cutting and pasting techniques. It has to be said that, at least on a narrow interpretation, this programme did not yield as much fruit as one might have hoped, and its goals have been to a large extent overtaken by events. The main lines of progress in this area (aside from algebro-geometrical techniques) came roughly thus. Firstly, through work of Mrowka and others involving cutting and pasting along 3-tori which, while it probably could be incorporated in a suitable generalisation of Floer theory, was not formulated explicitly in these terms. Secondly, through work of Kronheimer and Mrowka using singular connections (although again a version of Floer theory appeared in their arguments). Thirdly, and most decisively, through the introduction in 1994 of the Seiberg–Witten invariants. Leaving aside the well-known issue of the equivalence of the two theories, this last gives a more economical and powerful basis for the entire subject and makes the older instanton theory largely redundant as far as applications to 4-manifold topology go.

While it cannot be denied that the material in this book is less topical now than a decade ago (and at some points the text may have a slightly dated air, reflecting the long period over which it has been written) the author hopes that it is still worthwhile to present this material. We mention three grounds for this hope. First, the main thrust of the first part of the book is to develop certain differential-geometric and analytical techniques which apply to a wide range of problems, going beyond Yang–Mills theory (for example to the analogous symplectic Floer theory, to the Seiberg–Witten version of the Floer theory, to gluing problems for other structures such as self-dual metrics and metrics of special holonomy). Second, Floer's fundamental idea of defining 'middle-dimensional homology' for suitable infinite-dimensional manifolds is such an appealing one, and again one which in principle could appear in many different contexts, that it seems to deserve a thorough treatment. Third, while, as we have said above, some of the original motivation for the theory *vis-à-vis* 4-manifold topology is now reduced, there are intriguing questions which remain to be settled in setting up the Floer theory and understanding the whole relation between the instanton invariants and the Seiberg–Witten invariants. Some of these, in particular the appearance of modular forms, are touched on in Chapter 8. The Seiberg–Witten version of the Floer theory is a topic which is being very actively developed at the time of writing and, in conjunction with Floer's original groups, is expected to have important consequences in 3-manifold theory.

There are many topics omitted from this book. (In some cases these are things which we had hoped to include, in earlier and more ambitious plans, but found the energy wanting when it came to the point.) There are absolutely no examples: this is an entirely 'theoretical' treatment. We do not discuss the Casson invariant of homology spheres [46], or Floer's exact surgery sequence [8]. We do not mention Fukaya's extension of Floer's homology groups [9]. We do not have anything to say explicitly about the related theories developed by Taubes [47] and Morgan, Mrowka and Rubermann [36]. We do not say anything about the various interesting links between Floer's theory and the moduli spaces of flat connections over surfaces, and with algebraic geometry. Finally we say nothing about many of the deeper and more recent developments, connected with the Seiberg–Witten theory, such as the work of Muñoz [37] and Froyshov [25] on the 'finite type' condition. We do not discuss the Seiberg–Witten equations [51], [13], and the variant of the Floer theory they define. Except for the discussion of Fintushel and Stern's work of 1993 in Chapter 8 we have confined ourselves to an exposition of ideas that were current *circa* 1990.

On the other hand, we do digress from the narrow goal of setting up the Floer theory at a number of points. Thus, for example, we develop some of the main analytical results (in Chapter 4) in more generality than we need, because the ideas seem interesting and useful in other applications. We attempt to say a little about the background in mathematical physics, and the analogy with the symplectic theory.

The general scheme of the book is as follows. The first part (Chapters 2–5) aims to give a complete definition of the Floer groups of a homology 3-sphere: essentially following Floer's original paper. Chapter 6 develops the basic connection with 4-manifold invariants. The thrust of the first part is towards the geometrical and analytical techniques: at the beginning of Chapter 6 we step back to discuss the overall conceptual picture. Some readers may wish to look at the beginning of Chapter 6 at an earlier stage. Chapter 7 is devoted to refinements of the theory, mainly involving ideas from algebraic topology. This sets the stage for Chapter 8 in which, as we have mentioned, we discuss open problems and likely further developments.

The author expresses thanks to all the colleagues who have provided both help with this work and encouragement to complete the task. Two people should be mentioned in particular. Dieter Kotschick made notes of the original course of seminars which was the starting point for this project, and provided a great deal of help with the early development of the manuscript. Mikio Furuta was a participant in the seminar and contributed some invaluable ideas and drafts covering the less standard material in Chapter 7. In particular the formulation of the 'category of chain complexes' in that Chapter is due to Furuta.

2

Basic material

2.1 Yang–Mills theory over compact manifolds

In this Section we recall the rudiments of Yang–Mills theory in the standard situation – treated in numerous references – of a compact base manifold. (In general in this Chapter we follow the notation of [17].) So let V be a compact, connected, smooth manifold of dimension n, G be a compact Lie group and $P \to V$ be a principal G bundle over V. The gauge group $\mathcal{G}$ of automorphisms of P, covering the identity on V, acts on the space $\mathcal{A}$ of all connections on P by

$$g(A) = A - (d_A g)g^{-1}.$$

In Yang–Mills theory one needs to work with connections modulo gauge equivalence, i.e. modulo the action of $\mathcal{G}$, and to do this one can form the quotient spaces $\mathcal{B} = \mathcal{A}/\mathcal{G}$. This quotient is made more complicated by the possible existence of reducible connections, by which we mean connections A whose stabilisers Γ_A in $\mathcal{G}$ are larger than the centre $C(G)$ of G. (The stabiliser Γ_A is always a compact Lie group – the centraliser of the holonomy group of the connection A.) To avoid these complications one can restrict to the subset

$$\mathcal{B}^* = \{[A] \in \mathcal{B} : \Gamma_A \cong C(G)\}.$$

This is an open, dense subset of $\mathcal{B}$ (so long as the dimension n is greater than 1). We can make $\mathcal{B}^*$ into a smooth infinite-dimensional Banach manifold if we complete our spaces in suitable Sobolev norms. For example, we can take connections of class L^p_{k-1}, acted on by gauge transformations of class L^p_k (i.e. k derivatives in L^p). If the indices k and p satisfy the inequality $k - (n/p) > 0$ the L^p_k gauge transformations are *continuous* and the completion is naturally a Banach Lie group.

For the rest of this Chapter we shall denote by $\mathcal{G}$ and $\mathcal{A}$ these Sobolev completions. Thus $\mathcal{G}$ is a Banach Lie group acting smoothly on the Banach manifold $\mathcal{A}$.

To see the manifold structure of $\mathcal{B}^*$ explicitly we have to find slices for the action of $\mathcal{G}$. Fixing a background connection A_0 we have

$$\mathcal{A} = A_0 + \Omega^1(\mathfrak{g}_P)$$

where $\mathfrak{g}_P$ is the bundle of Lie algebras associated to P by the adjoint action of G. The tangent space to the orbit $\mathcal{G}(A_0)$ at A_0 is the image of the covariant derivative

$$d_{A_0} : \Omega^0(\mathfrak{g}_P) \to \Omega^1(\mathfrak{g}_P).$$

If V is equipped with a Riemannian metric then the space of connections becomes an infinite-dimensional, affine, Euclidean space, with the $\mathcal{G}$-invariant metric inherited from the standard L^2 metric on $\Omega^1(\mathfrak{g}_P)$. (This is not, of course, the same as an L^p_{k-1} metric used in completing $\mathcal{A}$.) There is then a standard choice of complementary subspace, namely the L^2 orthogonal complement. By Hodge theory, this is just the kernel of the formal adjoint operator

$$d^*_{A_0} : \Omega^1(\mathfrak{g}_P) \to \Omega^0(\mathfrak{g}_P).$$

The set of connections $A_0 + a$, for small a and with $d^*_{A_0} a = 0$, forms a local slice for the action of $\mathcal{G}$, and these slices give charts for $\mathcal{B}^*$. At the linear level we can identify the tangent space

$$T_{[A_0]}\mathcal{B}^* = \Omega^1(\mathfrak{g}_P)/\operatorname{Im} d_{A_0} = \ker d^*_{A_0}. \tag{2.1}$$

The *curvature* F_A or $F(A)$ of an L^p_{k-1} connection lies in L^p_{k-2}, so long as the inequality $k > n/p$ holds. The curvature can be regarded as a $\mathcal{G}$-equivariant map

$$F : \mathcal{A} \to \Omega^2(\mathfrak{g}_P),$$

whose derivative at a connection A_0 is the coupled exterior derivative

$$d_{A_0} : \Omega^1(\mathfrak{g}_P) \to \Omega^2(\mathfrak{g}_P).$$

This is obtained by linearising the formula

$$F_{A+a} = F_A + d_A a + a \wedge a. \tag{2.2}$$

Down on $\mathcal{B}^*$ we can think of the curvature as a section of a bundle of Banach spaces, the bundle over $\mathcal{B}^*$ associated to the action of $\mathcal{G}$ on $\Omega^2(\mathfrak{g}_P)$.

2.2 The case of a compact 4-manifold

Now we specialise to the case when $V = X^4$ is an oriented Riemannian 4-manifold. On X we have the Hodge $*$-operator, which acts on (bundle-valued) 2-forms, with square 1. Decomposing the curvature F_A of a connection A according to the decomposition $\Omega^2 = \Omega^+ \oplus \Omega^-$ of the 2-forms into *self-dual* and *anti-self-dual* parts (the ± 1 eigenspaces of $*$) we write $F_A = F_A^+ + F_A^- \in \Omega^+(\mathfrak{g}_P) \oplus \Omega^-(\mathfrak{g}_P)$. The *instanton* or *anti-self-dual (ASD) Yang–Mills* equation for a connection over any oriented Riemannian 4-manifold is the equation

$$F^+(A) = 0.$$

We refer to the solutions as 'instantons' or 'ASD connections'. Note that the instanton equation is *conformally invariant.*

The linearisation of the ASD equation about a given solution A is obtained by taking the self-dual part of Equation 2.2. We have

$$F_{A+a}^+ = d_A^+ a + (a \wedge a)^+,$$

where d_A^+ is the projection of the exterior derivative to $\Omega^+(\mathfrak{g}_P)$. We get a complex

$$\Omega^0(\mathfrak{g}_P) \xrightarrow{d_A} \Omega^1(\mathfrak{g}_P) \xrightarrow{d_A^+} \Omega^+(\mathfrak{g}_P). \tag{2.3}$$

Notice that these operators are defined for any connection A over X, not just the instantons, and in general the composite $d_A^+ \circ d_A$ is given by the algebraic action of F_A^+.

Plainly the linearisation of the instanton equation is the equation, for $a \in \Omega^1(\mathfrak{g}_P)$,

$$d_A^+ a = 0.$$

The instanton equation is gauge-invariant, so to study the solutions near A we may as well restrict to the slice defined by Equation 2.1. Thus the linearised equation modulo gauge equivalence can be written as the single equation

$$D_A(a) = 0, \tag{2.4}$$

where $D_A = -d_A^* \oplus d_A^+ : \Omega^1(\mathfrak{g}_P) \to \Omega^0(\mathfrak{g}_P) \oplus \Omega^+(\mathfrak{g}_P)$. The operator d_A^+ is not elliptic by itself but the gauge-fixing condition built into D_A makes this latter operator elliptic. (Thus the instanton equation, viewed modulo gauge transformations, is a non-linear elliptic PDE.) Like any elliptic operator over a compact manifold, D_A has a Fredholm index:

$$\operatorname{ind} D_A = \dim\ker D_A - \dim\ker D_A^*.$$

The Atiyah–Singer index theorem gives a topological formula for this index which takes the form

$$\operatorname{ind} D_A = c(G)\kappa(P) - \dim G(1 - b_1 + b^+).$$

Here $c(G)$ is a normalising constant, $\kappa(P)$ is a characteristic number of P obtained by evaluating a 4-dimensional characteristic class on the fundamental cycle $[X]$, b_1 is the first Betti number of X and b^+ is the rank of a maximal positive subspace for the *intersection form* on $H_2(X)$. We now focus on the case when $G = SU(2)$ and we can take κ to be the second Chern class $c_2(P)$. Then the index formula becomes

$$8c_2(P) - 3(1 - b_1 + b^+). \tag{2.5}$$

(In Chapter 5 we will discuss the case of $U(2)$ and $SO(3)$ connections.)

Chern–Weil theory expresses the topological characteristic number $\kappa(P)$ as a curvature integral. Specialising again to the case of the group $SU(2)$ where $\kappa = c_2$ we have

$$\kappa(P) = \frac{1}{8\pi^2}\int_X \operatorname{Tr}(F_A^2). \tag{2.6}$$

This applies, of course, to any connection A on P. The wedge product form is equal to the square of the norm on self-dual 2-forms and opposite on the anti-self-dual forms, so we have the fundamental equation

$$\kappa(P) = \frac{1}{8\pi^2}\int_X \left(\left|F_A^-\right|^2 - \left|F_A^+\right|^2\right) d\mu. \tag{2.7}$$

So a connection is an instanton if and only if

$$\kappa(P) = \frac{1}{8\pi^2}\int_X |F|^2 \, d\mu. \tag{2.8}$$

This shows, in particular, that $\kappa(P) \geq 0$ if P supports an ASD connection. (And if $\kappa(P) = 0$ the connection must be *flat* – associated to a representation of $\pi_1(X)$.)

2.3 Technical results

We will now recall briefly the main theorems about Yang–Mills instantons, from the point of view of applications to 4-manifold differential topology. These will be used in Chapters 4 and 5 when we extend the theory to certain non-compact base manifolds. We refer to [17] for proofs.

The analytical theory of instantons is underpinned by two basic theorems of Uhlenbeck. These are, in the first instance, local results: they deal with connections over the 4-ball B^4 and the punctured 4-ball $B^4 \setminus \{0\}$ respectively, and the hypotheses involve the curvature integral

$$\|F\|^2 \equiv \int |F|^2 \, d\mu,$$

which we have encountered already in the Chern–Weil theory for instantons above. Note that, like the instanton equation, this integral is conformally invariant – it depends only on the conformal class of the Riemannian metric. The first theorem asserts that there is a constant ϵ such that any connection A over B^4 with $\|F(A)\|^2 < \epsilon$ can be represented by a connection matrix which satisfies the Coulomb gauge-fixing condition, and whose L^2_1 norm is estimated by the L^2 norm of F. If the connection also satisfies the ASD equations then one obtains interior estimates on all higher derivatives of the connection matrix. The Ascoli–Arzelà theorem then implies that any sequence of such ASD connections, with small curvature, has a convergent subsequence. The second theorem, Uhlenbeck's 'removable singularities theorem', asserts that any ASD connection over the punctured ball whose curvature is in L^2 can be extended smoothly over the puncture. Strictly, this also involves extending the bundle on which the original connection is defined. These two results, together with some 'patching' arguments, lead to a global compactness principle which we will use in the following form.

Proposition 2.1 *Let U be an oriented Riemannian 4-manifold, possibly non-compact, and A_α an infinite sequence of ASD connections on a bundle P over U. If the curvatures satisfy a fixed L^2 bound $\|F(A_\alpha)\| \leq C$, for some $C \in \mathbf{R}$, then there is a subsequence $\{\alpha'\}$ such that the connections $A_{\alpha'}$ converge as $\alpha' \to \infty$ in the following sense. There are points $x_1, \ldots, x_l \in U$ (not necessarily distinct), with $l \leq C/8\pi^2$, an ASD connection A on a bundle Q over U and bundle isomorphisms*

$$\rho_{\alpha'} : Q|_{U \setminus \{x_1, \ldots, x_l\}} \to P|_{U \setminus \{x_1, \ldots, x_l\}}$$

such that

(1) *$\rho^*_{\alpha'}(A_{\alpha'})$ converges to A, uniformly with all derivatives on compact subsets of the punctured manifold,*

(2) *the curvature densities* $|F(A_{\alpha'})|^2$ *converge to*

$$|F(A)|^2 + 8\pi^2 \sum_{i=1}^{l} \delta_{x_i},$$

in the sense that for any compactly supported, continuous, function f *on* U *we have*

$$\int_U f\ |F(A_{\alpha'})|^2\, d\mu \to \int_U f\ |F(A)|^2\, d\mu + 8\pi^2 \sum f(x_i).$$

We shall refer to this kind of convergence as 'weak convergence'. The fact that one only gets over the punctured manifold is the famous 'bubbling phenomenon'. Notice that (2) implies that

$$\int_U |F(A)|^2 + 8\pi^2 l \leq C,$$

and if U is compact equality must hold. Also, if U is compact, the hypotheses on the curvature are automatically satisfied by the Chern–Weil equality Equation 2.8. Then the last part of the Proposition implies that

$$\kappa(Q) \leq \kappa(P) + l.$$

The next piece of theory we need refines our discussion in Section 2.2 of the linearisation of the ASD equations. For a bundle P over a compact oriented 4-manifold X, with Riemannian metric g, we define the *moduli space*, $M_P(g)$, of instantons on P to be the set of solutions of the ASD equation modulo gauge equivalence. Thus $M_P(g)$ is a subset of the quotient $\mathcal{B} = \mathcal{A}/\mathcal{G}$ of *all* connections modulo equivalence. On the other hand, we have attached an integer to P, the index of the operator D_A, for any connection A on P. This is the 'virtual dimension' of the moduli space $M_P(g)$, in that roughly speaking $M_P(g)$ is typically a manifold of this dimension. More precisely, we have

Proposition 2.2 (1) *If the structure group of* P *is* $SU(2)$ *and* $\kappa(P) > 0$ *then for generic metrics* g *on* X *the intersection* $M_P^*(g)$ *of the moduli space* $M_P(g)$ *with* $\mathcal{B}^*$ *(the 'irreducible' connections) is a smooth manifold of dimension* $\operatorname{ind} D_A$. *For each* $[A] \in M_P^*(g)$ *the tangent space to the moduli space is naturally identified with the kernel of* D_A *and the cokernel of this operator is zero.*

(2) *If, in addition, the 4-manifold X satisfies the condition $b^+(X) > 0$ (i.e. the intersection form is not negative definite) then, for generic metrics on X, $M^*_P(g) = M_P(g)$.*

In Proposition 2 we mean by 'generic metrics' the metrics in a second category subset of the space of C^k metrics for some fixed $k > 2$. (It may reassure the reader to know that for all practical purposes one can work with an open dense subset of the smooth metrics, or even real analytic metrics.) The third topic we wish to mention is the orientation of the moduli spaces. For a generic metric as above the $SU(2)$ moduli spaces are orientable manifolds. The discussion of the orientations is in reality completely independent of the moduli space: it is a part of index theory dealing with a *determinant line bundle*, a real line bundle $\lambda \to \mathcal{B}$, whose fibres are

$$\lambda_{[A]} = \Lambda^{\max} \ker D_A \otimes (\Lambda^{\max} \ker D_A^*)^*.$$

There is a canonical isomorphism between the restriction of λ to $M^*_P(g)$ and the orientation line bundle of the moduli space, so a trivialisation of λ induces an orientation of the moduli space.

Proposition 2.3 *The line bundle λ is trivial over $\mathcal{B}$. A trivialisation of λ can be specified canonically by an orientation of $H^1(X;\mathbf{R}) \oplus H^+$ where $H^+ \subset H^2(X;\mathbf{R})$ is a maximal positive subspace.*

2.4 Manifolds with tubular ends

In this book we will study moduli spaces of instantons over non-compact 4-manifolds. For any complete oriented Riemannian 4-manifold X we get natural moduli problems – to study connections A on bundles over X with finite 'energy', i.e.

$$\int_X |F(A)|^2 \, d\mu < \infty,$$

modulo gauge equivalence. The naturality of this class of connections is suggested by the importance of the L^2 norm of the curvature throughout the whole theory. In this direction there are certainly interesting analytical questions which can be asked about rather general classes of non-compact 4-manifolds, but in this book we shall be considering only manifolds of a very simple, standard, form.

The manifolds we will consider are 4-*manifolds with tubular ends*. Such a manifold is a complete Riemannian 4-manifold X which contains a finite number of 'ends' U_i, $i = 1, \ldots, n$: open subsets of X isometric to half-tubes $Y_i \times (0, \infty)$, where Y_i are compact Riemannian 3-manifolds (which we will refer to as the 'cross-sections' of the tubes). Thus the closure $\overline{X \setminus \bigcup_{i=1}^{n} U_i}$ is a compact manifold-with-boundary. We suppose all our manifolds are oriented, and for definiteness we assume that the orientations at the ends are given by

$$dt \wedge dy_1 \wedge dy_2 \wedge dy_3,$$

where dt is the standard 1-form on $\mathbf{R}$ and $dy_1\, dy_2\, dy_3$ defines the orientation of the cross-section.

The half-tube $S^3 \times (0, \infty)$ (where the 3-sphere has the standard round metric) is conformally equivalent to the punctured ball $B^4 \setminus \{0\}$, with the Euclidean metric. A conformal equivalence is given, with respect to 'generalised polar co-ordinates' (r, θ) on the punctured ball, by the map

$$(t, \theta) \to (e^{-t}, \theta).$$

This means that a manifold X with tubular ends, whose ends each have the 3-sphere as cross-section, is conformally equivalent to a punctured manifold $\tilde{X} \setminus \{p_1, \ldots, p_n\}$, where $\tilde{X}$ is compact. Both the instanton equation and the L^2 norm of the curvature are conformally invariant so the moduli problem for X is equivalent to that on the punctured manifold. In turn, by Uhlenbeck's removable singularities theorem, this is equivalent to the moduli problem for the compact manifold $\tilde{X}$. Thus it is natural to expect that the theory for manifolds with general tubular ends will have many of the same basic features as the theory for compact manifolds, and in the subsequent Sections we will see that this is indeed the case.

2.5 Yang–Mills theory and 3-manifolds

2.5.1 Initial discussion

We now switch to three dimensions and fix attention on a compact oriented 3-manifold Y. The main theme of this Section is to see the interaction between gauge theory in three dimensions (on Y) and four dimensions (in the fashion considered above), specifically on the *tube* $Y \times \mathbf{R}$. The main idea is that one can pass from connections on $Y \times \mathbf{R}$ to *one-parameter families* of connections over Y. In a sense, this is rather an elementary thing. We can explain the main point first in a simple way, using local trivialisations and co-ordinates. Let t be the standard parameter on the factor $\mathbf{R}$ in $Y \times \mathbf{R}$ and let y_i $(i = 1, 2, 3)$ be local co-ordinates on a patch in Y. (It is, of course, natural to have in mind a situation where $Y \times \mathbf{R}$ is 'space-time', and t appears as the time co-ordinate.) A connection over the tube is given locally by a connection matrix:

$$\mathbf{A} = A_0 \, dt + \sum_{i=1}^{3} A_i \, dy_i,$$

where A_0 and the A_i depend on all four variables t, y_1, y_2, y_3. We get a one-parameter family of connection matrices over an open set in Y in an obvious way, by discarding A_0 and considering t as an additional parameter. (As a point of notation, throughout this book, when we are discussing connections over 4-manifolds and 3-manifolds at the same time we will often use bold face for the former, but we will not do always do this, and we will revert to plain symbols when no confusion seems likely.) Slightly less obvious is the interaction between this construction and the notion of gauge equivalence of connections. For any given connection over the tube we can use parallel transport along the '$\mathbf{R}$-factor' to choose a connection matrix with $A_0 = 0$ (sometimes called a *temporal gauge*). Thus, in this gauge, we do not lose any information by discarding A_0. In this situation the curvature in a mixed ty_i-plane is given by the simple formula

$$F_{0i} = \frac{\partial A_i}{\partial t}. \tag{2.9}$$

Now let Y have a Riemannian metric with $*$-operator $*_3$. If ϕ is a 1-form on Y then, for the 4-dimensional $*$-operator defined with respect to the product metric on $Y \times \mathbf{R}$,

$$*(dt \wedge \phi) = *_3 \phi. \tag{2.10}$$

Thus the anti-self-dual forms are just those of the shape $\phi \wedge dt + *_3\phi$, and the instanton equation in a temporal gauge has the form of an 'evolution' equation for a one-parameter family $A(t)$ over Y:

$$\frac{\partial A(t)}{\partial t} = *_3 F(A(t)). \tag{2.11}$$

The observation that the instanton equation can be cast in this form lies at the root of Floer's approach, as we shall see.

2.5.2 The Chern–Simons functional

We will now go over these ideas again, taking a more systematic and invariant approach. We begin with bundle theory. If P is a bundle over the 3-manifold Y, and u is an automorphism of P (covering the identity) we can form a bundle $\mathbf{P}_u$ over $Y \times S^1$ by using u to glue the ends of $\pi_1^*(P)$ over $Y \times [0,1]$. It is an elementary fact that this sets up a one-to-one correspondence between isomorphism classes of bundles over $Y \times S^1$, isomorphic to P on $Y \times \{\text{pt.}\}$, and the *connected components* of the gauge group $\mathcal{G}_P$ of P on Y. On the other hand, if we choose a connection $\mathbf{A}$ on the bundle $\mathbf{P}_u$ over $Y \times S^1$ we get, in the manner above, a one-parameter family of gauge equivalence classes of connections over Y, giving us a map

$$\gamma_{u,\mathbf{A}} : S^1 \to \mathcal{B}_P.$$

This cannot, in general, be lifted to a loop in $\mathcal{A}_P$. Indeed, if we cut open the circle and lift the resulting map of $[0,1]$ to $\mathcal{A}_P$, the images in $\mathcal{A}_P$ of the end points differ precisely by the action of the automorphism u. If we assume that the structure group G is connected and restrict to the (contractible) open subset $\mathcal{A}_P^*$ of irreducible connections we get in this way a natural isomorphism between $\pi_1(\mathcal{B}_P^*)$ and the group of components $\pi_0(\mathcal{G}_P)$. In sum, then, we see that the fundamental group of the space $\mathcal{B}_P^*$ is isomorphic to the set of equivalence classes of bundles on $Y \times S^1$. In particular, the 4-dimensional characteristic class over the product detects essential loops in the connection space. It is easy to see that these loops remain essential if we include the reducible connections, so we get a homomorphism

$$\kappa : \pi_1(\mathcal{B}_P) \to \mathbf{Q}, \tag{2.12}$$

i.e. a class $\hat{\kappa}$ in $H^1(\mathcal{B}_P; \mathbf{Q})$.

Let us now specialise to bundles with gauge group $G = SU(2)$. In this case the bundle P over Y is necessarily trivial and we get an isomorphism

$$c_2 : \pi_1(\mathcal{B}_P) \to \mathbf{Z}.$$

We will now take a more differential-geometric approach to these matters, by introducing the *Chern–Simons function* which lies at the heart of the Floer theory. This is a map

$$\vartheta : \mathcal{B}_P \to \mathbf{R}/\mathbf{Z}, \tag{2.13}$$

which can be defined in various ways. We will describe three slightly different approaches: the equivalence between these is not a matter of any difficulty, but the different points of view are useful. For simplicity we will work in the case of $SU(2)$ bundles

First definition This follows a formal pattern which will recur several times in different contexts in this book (for example in the discussion of orientation in Chapter 5). Choose an oriented 4-manifold X' with boundary Y, and an extension of the bundle P over X'. (Here we use the triviality of the cobordism group in three dimensions.) Then, for a given connection A on P, choose an extension $\mathbf{A}'$ over X' and set

$$\vartheta(A) = \frac{1}{8\pi^2}\int_{X'} \mathrm{Tr}\big(F_{\mathbf{A}'}^2\big). \tag{2.14}$$

The key point is that this is the integral which, in the case of a closed base manifold, would give the characteristic number c_2. If X'' and $\mathbf{A}''$ are different choices we form a connection $\mathbf{A}$ on a closed 4-manifold

$$X = X' \cup_Y X''$$

by gluing over Y. Then the integrality of the Chern–Weil form over the closed manifold (Equation 2.6) implies that, modulo $\mathbf{Z}$, the two choices give the same value for ϑ. (The connection $\tilde{A}$ may not be smooth, but a moment's thought shows that this does not affect the argument.) The definition is manifestly gauge-invariant, and so defines a map as in Formula 2.13. This has a number of more or less formal properties, deriving from the nature of the definition. First, ϑ is 0 on the trivial connection, since we can make a trivial extension over X. Second, if Z is a 4-manifold with two boundary components $\overline{Y}$ and Y' and $\mathbf{A}$ is a connection on a bundle over Z which restricts to A and A' over the boundaries we have

$$\vartheta(A') = \vartheta(A) + \frac{1}{8\pi^2}\int_Z \mathrm{Tr}\big(F_{\mathbf{A}}^2\big) \bmod \mathbf{Z}. \tag{2.15}$$

This follows from the obvious gluing argument.

Third, consider a connection $\mathbf{A}$ on a bundle $\mathbf{P}$ over $S^1 \times Y$. The Chern–Simons function gives us a map

$$\theta : S^1 \to \mathbf{R}/\mathbf{Z} \equiv S^1,$$

with $\theta(t) = \vartheta(\mathbf{A}|_{\{t\}\times Y})$. We claim that the *degree* of this map is just the Chern number $c_2(\mathbf{P})$. This is clear from the Chern–Weil formula for c_2, since as a special case of Equation 2.15 we have, for any t and t',

$$\theta(t') - \theta(t) = \frac{1}{8\pi^2} \int_{[t,t']\times Y} \mathrm{Tr}(F_{\mathbf{A}}^2) \mod \mathbf{Z},$$

so the map $\tilde{\theta} : [0,1] \to \mathbf{R}$,

$$\tilde{\theta}(t) = \frac{1}{8\pi^2} \int_{[0,t]\times Y} \mathrm{Tr}(F_{\mathbf{A}}^2),$$

gives a lift of θ, and hence the degree of θ is $\tilde{\theta}(1) = c_2(\mathbf{P})$.

Second definition Here we work directly on the 3-manifold. We choose a trivialisation, so that a connection becomes identified with a Lie algebra-valued 1-form, and set

$$\vartheta'(A) = \frac{1}{8\pi^2} \int_Y \mathrm{Tr}\left(A \wedge dA + \frac{2}{3} A \wedge A \wedge A \right). \tag{2.16}$$

This turns out to depend on the trivialisation only up to an integer. We can see this in a way which ties in with our first definition, by choosing a connection $\mathbf{A}$ on the bundle $\pi^*(P)$ over $Y \times [0,1]$ which agrees with A on $Y \times \{1\}$ and is the trivial product connection over $Y \times \{0\}$. Then, as a special case of Equation 2.15, we have

$$\vartheta(A) = \frac{1}{8\pi^2} \int_{Y\times[0,1]} \mathrm{Tr}(F(\mathbf{A})^2).$$

On the other hand, in the trivialisation of our bundle over $Y \times [0,1]$, the connection is given by a connection form $\mathbf{A}$ and an easy calculation gives the identity of 4-forms on $Y \times [0,1]$:

$$d\left(\mathrm{Tr}\left(\mathbf{A} \wedge d\mathbf{A} + \frac{2}{3} \mathbf{A} \wedge \mathbf{A} \wedge \mathbf{A} \right) \right) = \mathrm{Tr}(F(\mathbf{A})^2). \tag{2.17}$$

Now the equality of ϑ and ϑ', modulo integers, follows from Stokes' theorem, since

$$\int_{Y\times[0,1]} \mathrm{Tr}(F(\mathbf{A})^2) = \int_Y \mathrm{Tr}\left(A \wedge dA + \frac{2}{3} A \wedge A \wedge A \right).$$

One might object that this definition relies on our ability to 'spot' the expression in Equation 2.16 which satisfies Equation 2.17. To derive the expression systematically one can consider an explicit choice of connection $\mathbf{A}$, given in a trivialisation by

$$\mathbf{A}_{t,y} = t A_y.$$

Then one can perform the integral in Equation 2.17 explicitly. It is useful to have also the formula

$$\vartheta(A+a) = \vartheta(A) + \frac{1}{8\pi^2}\int_Y \operatorname{Tr}\left(d_A a \wedge a + \frac{2}{3} a \wedge a \wedge a\right) \tag{2.18}$$

which is a simple generalisation of Equation 2.16.

We mention here that the Chern–Simons theory itself is not in any way special to three dimensions: there are analogous invariants in any odd dimension associated to the even-dimensional Chern–Weil integrands – the prototype is the case of connections with group S^1 and with S^1 as the base manifold. Then the analogous invariant, associated to the first Chern class, is just the holonomy of a connection around the circle. For the general theory of these 'secondary characteristic classes', which can be developed more systematically by working on the total space of the principal bundle, see [10].

Third definition A third approach to the definition of the Chern–Simons function takes the point of view of differential geometry on the infinite-dimensional spaces $\mathcal{A}_P, \mathcal{B}_P$ of connections (although the infinite-dimensionality is not at all significant here, since one can always restrict to finite-dimensional subsets). We consider the 1-form ρ on $\mathcal{A}_P$ which assigns to a tangent vector $a \in \Omega^1(\mathfrak{g}_P)$ at a point $A \in \mathcal{A}_P$ the number

$$\rho_A(a) = \int_Y \operatorname{Tr}(F_A \wedge a). \tag{2.19}$$

Then we have

$$\rho_{A+b}(a) - \rho_A(a) = \int_Y \operatorname{Tr}(d_A b \wedge a) + \mathrm{O}(b)^2,$$

by Equation 2.2. Now

$$\int_Y \operatorname{Tr}(d_A b \wedge a - b \wedge d_A a) = \int_Y d\big(\operatorname{Tr}(b \wedge a)\big) = 0,$$

by Stokes' theorem. Thus the leading term in Equation 2.19 is symmetric in a, b. This is precisely what it means for ρ to be a *closed* 1-form. It follows that ρ can be written as the derivative of a function $8\pi^2\vartheta''$ on the contractible space $\mathcal{A}_P$. Now, by a similar application of Stokes' theorem one sees that ρ vanishes on tangent vectors $a = d_A\xi$ along the $\mathcal{G}$ orbits, so ρ descends to a form on the quotient $\mathcal{B}_P^*$, and ϑ'' descends at least locally to $\mathcal{B}_P$. More precisely, it descends to define a map into $\mathbf{R}$ modulo a homomorphic image of $H_1(\mathcal{B}_P)$ given by the 'periods' of the 1-form. It is then easy to see, by linearising Equation 2.18 about $a = 0$, that when suitably normalised this is just the Chern–Simons map. More precisely, the derivative of ϑ is $(8\pi^2)^{-1}\rho$.

The preceding discussion shows that *the critical points of the Chern–Simons function on $\mathcal{B}_P$ are the gauge equivalence classes of flat connections on P.* This is one of the fundamental facts underpinning Floer's theory. The holonomy of a connection sets up a one-to-one correspondence between the flat connections (modulo gauge equivalence) and the representations of the fundamental group $\pi_1(Y)$ in $SU(2)$ (modulo conjugation), and thus depends only on the topology of the 3-manifold. Perhaps somewhat surprisingly, we will make almost no use of this basic fact in this book, but it is obviously of great importance in understanding the relation between the differential-geometric constructions and manifold topology. We write $\mathcal{R}_Y$ for the set of equivalence classes of flat connections over Y, and $\mathcal{R}_Y^*$ for the equivalence classes of irreducible flat connections.

2.5.3 The instanton equation

We now turn to the instanton equations. First we should make the relation between connections over the tube and one-parameter families of connections over Y more precise. As in Subsection 2.5.2, a connection $\mathbf{A}$ over $Y \times \mathbf{R}$ defines, by restricting to slices $Y \times \{\text{pt.}\}$, a map $\gamma_{\mathbf{A}} : \mathbf{R} \to \mathcal{B}_P$ and this is plainly gauge-invariant. Conversely, suppose γ is a path in $\mathcal{B}_P$, and $\tilde{\gamma} : \mathbf{R} \to \mathcal{A}_P$ is a lift. Write $A_t = \tilde{\gamma}(t)$. Then we can define a connection in four dimensions, on the bundle $\pi^*(P)$, in the familiar way

$$\mathbf{A}_{t,y} = A(t)_y.$$

This is an inverse to the preceding construction, in that $\gamma_{\mathbf{A}} = \gamma$, but the connection $\mathbf{A}$ depends on the choice of lift $\tilde{\gamma}$. Different choices of lift will give connections over the tube which are *not* gauge-equivalent. We can get a precise correspondence in two ways. On the one hand we can work with paths $\tilde{\gamma}$ in $\mathcal{A}_P$. The constructions above show that there is a natural one-to-one correspondence:

$$\mathcal{B}_{\pi^*(P), Y \times \mathbf{R}} \leftrightarrow \frac{\mathrm{Maps}(\mathbf{R}, \mathcal{A}_P)}{\mathcal{G}_P}.$$

On the other hand we can make a canonical choice of a lift $\tilde{\gamma}$ of any path in $\mathcal{B}_P$. For this we use a Riemannian metric on Y and the Coulomb gauge slices considered in Section 2.1. We say that a lift $A_t = \tilde{\gamma}$ of γ is *horizontal* if

$$d^*_{A_t}\left(\frac{\partial A_t}{\partial t}\right) = 0.$$

Then any path in $\mathcal{B}_P$ has a horizontal lift, unique up to the action of $\mathcal{G}_P$. Away from reducible connections this assertion is a simple consequence of the existence and uniqueness of solutions to ordinary differential equations (see [17][Section 6.3.1]), and the ideas can be extended to reducibles (we need not worry about potential technical difficulties here since we need the result only for motivation). The horizontal condition can be expressed directly in terms of the connection in four dimensions. We say a connection $\mathbf{A}$ on $\pi^*(P)$ is in temporal gauge if the covariant derivative of any section lifted up from Y along the vector $\frac{d}{dt}$ in the $\mathbf{R}$-direction vanishes. For such a connection we have the invariant version of Equation 2.9:

$$\frac{\partial}{\partial t} A_t = i_t(F_{\mathbf{A}}),$$

where i_t denotes contraction by the vector $\frac{d}{dt}$, and A_t is the family of connections on P obtained by restriction to slices. It follows then that

$$d^*_{A_t}\left(\frac{d}{dt}(A_t)\right) = i_t(d^*_{\mathbf{A}} F_{\mathbf{A}}).$$

So the connections over the tube which give horizontal paths, viewed as one-parameter families, are those satisfying the equation

$$i_t(d^*_{\mathbf{A}} F_{\mathbf{A}}) = 0. \tag{2.20}$$

We obtain a natural one-to-one correspondence

$$\{[\mathbf{A}] \in \mathcal{B}_{\pi^*(P), Y\times\mathbf{R}} : i_t(d^*_{\mathbf{A}} F_{\mathbf{A}}) = 0\} \leftrightarrow \text{Maps}\,(\mathbf{R}, \mathcal{B}_P).$$

Now, turning to instantons, we recall first the well-known fact that the ASD equation $F_{\mathbf{A}} + *F_{\mathbf{A}} = 0$, together with Bianchi identity $d_{\mathbf{A}} F_{\mathbf{A}} = 0$, implies that any instanton over $Y \times \mathbf{R}$ satisfies the *Yang–Mills equation*

$$d^*_{\mathbf{A}} F_{\mathbf{A}} = 0,$$

and hence the horizontality condition Equation 2.20. So an instanton over the tube yields a horizontal path, and we can identify the instantons, up to gauge equivalence, with a subset of the paths in $\mathcal{B}_P$. By Equation 2.11 the condition for such a path γ to correspond to an instanton is just that it satisfies the differential equation

$$\frac{d}{dt} A_t = *_3 F(A_t).$$

This is the equation for the integral curves of a vector field $\mathcal{V}$ over the space of connections defined by

$$\mathcal{V}(A) = *_3 F(A).$$

We can interpret this as an equation for a path either in $\mathcal{A}_P$, or in the quotient: the two are equivalent since the vector field on $\mathcal{A}_P$ is $\mathcal{G}_P$-invariant and horizontal, because $d_A^* F_A = *d_A F_A = 0$ by the Bianchi identity. This is just the same observation linking horizontality to the instanton equation that we made above, viewed now from the 3-dimensional point of view.

We observe next that the vector field $\mathcal{V}$ is dual to the 1-form ρ we considered above, with respect to the L^2 metric on 1-forms. That is:

$$\rho_A(a) = \int_Y \mathrm{Tr}(F_A \wedge a) = \int_Y (*F_A, a) = \langle \mathcal{V}(A), a \rangle.$$

Again, we can work here either on $\mathcal{A}_P$ or on the quotient $\mathcal{B}_P$, using the metric on the tangent space of $\mathcal{B}_P$ induced from the L^2 metric by the horizontal slice.

We see then that the vector field $\mathcal{V}$ can be regarded as the *gradient* of a multiple of the Chern–Simons function (since $\rho = 8\pi^2 d\vartheta$). Of course this function only maps to the circle, but the gradient of such a map obviously makes sense. Thus we see that *the instanton equation over $Y \times \mathbf{R}$ can be interpreted as the gradient flow equation for the Chern–Simons function on $\mathcal{B}_P$.*

A useful example to have in mind is the well-known 1-instanton on $\mathbf{R}^4$, which has curvature density

$$|F| = \frac{1}{(1+r^2)^2}.$$

When viewed as a connection over $S^3 \times \mathbf{R}$, as in Section 2.4 above, we get a connection with curvature density

$$|F| = \frac{4}{\cosh^2(t)},$$

which, in the 3-dimensional picture, appears as a path which begins at the flat (trivial) connection at $t = -\infty$, winds once around the essential loop in the space of connections modulo equivalence, and returns to the same flat connection at $t = +\infty$.

2.5.4 Linear operators

We will now discuss the Chern–Simons function in more detail. We will consider the local behaviour of the function in a neighbourhood of a flat connection A, i.e. around a critical point, and we shall see that this naturally brings in the *twisted de Rham complex*

$$\Omega^0(\mathfrak{g}_P) \xrightarrow{d_A} \Omega^1(\mathfrak{g}_P) \xrightarrow{d_A} \Omega^2(\mathfrak{g}_P) \xrightarrow{d_A} \Omega^3(\mathfrak{g}_P), \tag{2.21}$$

defined by the flat connection. We denote the cohomology groups of this complex by H^i_A. As usual, the derivative Q of $\mathcal{V} = 8\pi^2 \operatorname{grad} \vartheta$ at A is intrinsically defined as a map from tangent vectors to tangent vectors; by Equation 2.2 it is given by

$$Q_A(a) = *d_A(a). \tag{2.22}$$

Notice that, again, $*d_A(a)$ is horizontal when A is flat so the discussion applies equally well to the quotient space. Thus we really want to consider the Hessian as an operator:

$$Q_A : \ker d_A^* \to \ker d_A^*.$$

This is, however, not very convenient. It is more practical to work with another operator which is essentially equivalent to the Hessian. For *any* connection A on P we let

$$L_A : \Omega^0_Y(\mathfrak{g}_P) \oplus \Omega^1_Y(\mathfrak{g}_P) \to \Omega^0_Y(\mathfrak{g}_P) \oplus \Omega^1_Y(\mathfrak{g}_P)$$

be given by the matrix of differential operators

$$\begin{pmatrix} 0 & -d_A^* \\ -d_A & *d_A \end{pmatrix}.$$

In classical vector calculus notation this is

$$\begin{pmatrix} 0 & -\operatorname{div} \\ -\operatorname{grad} & \operatorname{curl} \end{pmatrix}.$$

The operator L_A is self-adjoint, elliptic, and its square is the coupled Laplace operator on forms:

$$L_A^2 = \Delta_A = d_A^* d_A + d_A d_A^*.$$

Indeed, L_A can be thought of as 'one half' of the Euler characteristic operator $d_A + d_A^*$, mapping odd forms to even forms, when we identify 2-forms with 1-forms and 3-forms with 0-forms, using the $*$-operator on Y.

To relate L_A with the Hessian, we decompose the 1-forms as

$$\Omega^1(\mathfrak{g}_P) = \ker d_A^* \oplus \operatorname{Im} d_A.$$

Then, if A is flat,

$$L_A = Q_A \oplus S_A \tag{2.23}$$

where $S_A : \operatorname{Im} d_A \oplus \Omega^0(\mathfrak{g}_P) \to \operatorname{Im} d_A \oplus \Omega^0(\mathfrak{g}_P)$ is the operator

$$\begin{pmatrix} 0 & -d_A \\ -d_A^* & 0 \end{pmatrix}$$

and the square of S_A can be regarded as two copies of the Laplacian Δ_A on 0-forms, with the kernels of Δ_A in the two spaces being identified.

Now, as a self-adjoint elliptic operator, the spectrum of L_A is real and discrete. Our direct sum decomposition shows that it is the union of the spectra of Q_A and S_A. On the other hand, the eigenvalues of S_A are precisely the numbers of the form $\pm\sqrt{\lambda}$ where λ is an eigenvalue of the Laplacian on 0-forms. So we obtain the spectrum of Q_A by discarding these positive and negative square roots of the spectrum of Δ_A on $\Omega^0(\mathfrak{g}_P)$. The spectrum of Q_A also consists of square roots of the spectrum of the Laplacian on forms – this time on the kernel of d_A^* on 1-forms. However, it is important to note that we cannot predict from this Laplacian alone which sign of the square root will appear in the spectrum of Q_A.

We shall now describe the kernel of L_A in terms of the cohomology groups H_A^i of the de Rham complex. From the discussion above we see that the kernel of L_A decomposes into the direct sum of the kernel of $Q_A = *d_A$ in $\ker d_A^* \subset \Omega^1(\mathfrak{g}_P)$ and a copy of the kernel of d_A in $\Omega^0(\mathfrak{g}_P)$. The first of these is precisely the harmonic space

$$\ker d_A \cap \ker d_A^* \subset \Omega^1(\mathfrak{g}_P),$$

which is identified by the Hodge theory with the cohomology group H_A^1. The second space, the space of covariant constant sections of $\mathfrak{g}_P$, is precisely the cohomology group H_A^0. These two cohomology groups have a separate significance for the flat connection A. The cohomology H_A^0 is non-zero if and only if the connection is *reducible*, and the kernel is the Lie algebra of the isotropy group Γ_A of A in $\mathcal{G}_P$. (This isotropy group is ± 1 for an irreducible $SU(2)$ connection.) The cohomology H_A^1 is non-zero if and only if the connection can be deformed infinitesimally, within the space $\mathcal{B}_Y$ of gauge equivalence classes of flat connections on P – more precisely H_A^1 is the Zariski tangent space of this space.

These cohomology groups will play a very important role in setting up the theory in the next two Chapters, so we will fix on the following terminology.

Definition 2.4 *The flat connection A is*

- acyclic *if both cohomology groups H^0_A, H^1_A are zero,*
- non-degenerate *if H^1_A is zero (and H^0_A may or may not be zero).*

Notice that the higher cohomology groups of A are obtained by Poincaré duality from the first two, so the connection is acyclic if and only if the twisted de Rham complex is acyclic in the usual sense. The use of 'non-degenerate' corresponds to the standard definition of a non-degenerate critical point of a function, as in the Morse theory. We take the discussion further in Appendix A to this Chapter, where we discuss finite-dimensional models for the Chern–Simons functional around a flat connection.

We now return to the 4-dimensional point of view; and see how the linearisation in three dimensions discussed above ties up with those in four dimensions. Recall from Section 2.2 that in the usual instanton theory over a 4-manifold X one introduces a linear operator

$$D_{\mathbf{A}} = -d^*_{\mathbf{A}} \oplus d^+_{\mathbf{A}} : \Omega^1_X \to \Omega^0_X \oplus \Omega^+_X,$$

to describe the deformations of ASD connections. In the case when X is a Riemannian tube $Y \times \mathbf{R}$ we can identify the self-dual 2-forms with cotangent vectors on Y, using the map

$$\xi \to \xi \wedge dt + *_3\xi,$$

as in Equation 2.10, and similarly identify the 1-forms on X with $\Lambda^1_Y \oplus \Lambda^0_Y$ by taking the dt component in the obvious way. Thus if $\mathbf{A}$ is a connection $\pi^*(A)$ on a bundle $\pi^*(P)$ over the tube, we can regard $D_{\mathbf{A}}$ as a map from the space of sections over the tube of the pull-back bundle, which we just denote by $\Lambda^0_Y \oplus \Lambda^1_Y$, to itself. In these terms we can write the 4-dimensional covariant derivative $d_{\mathbf{A}}$ on Ω^0 as

$$d = \begin{pmatrix} \text{grad} \\ \frac{d}{dt} \end{pmatrix}$$

and $d^+_{\mathbf{A}}$ as

$$d^+_{\mathbf{A}} = \left(\frac{d}{dt}, \text{curl} - \text{grad}\right),$$

from which it follows that

$$D_{\mathbf{A}} = \frac{d}{dt} + L_A : \Gamma(\Lambda^0_Y \oplus \Lambda^1_Y) \to \Gamma(\Lambda^0_Y \oplus \Lambda^1_Y). \qquad (2.24)$$

Here the notation means that we take sections of the pull-back bundle over the tube, and we have suppressed the bundle of Lie algebras $\mathfrak{g}_P$. This formula will be central to the analysis of instantons on tubes in the following Chapters.

We have seen that the flat connections over Y can be viewed as the zeros of the vector field $\mathcal{V}$, the critical points of the Chern–Simons function, and the instantons on the tube are gradient flow lines of this vector field. This is of course reminiscent of a familiar picture in the calculus of variations where one characterises geometric objects as the critical points of a Lagrangian on some infinite-dimensional function space, for example geodesics and the energy function on the space of loops. In such a case also one has an associated evolution equation representing, at least formally, the gradient flow of the Lagrangian. There is however a radical difference between this instanton theory and such classical problems. In the classical theory the Hessian is typically represented by a second order operator of Laplace type, with a positive symbol. For example in the geodesic case this would be the Jacobi operator. The spectrum of this operator is then bounded below, with only finitely many negative eigenvalues. The number of negative eigenvalues yields the *index* of the critical point and, at least in favourable cases, the set of critical points and their indices is related, via the Morse theory, to the topology of the function space. Similarly the evolution equation is a parabolic equation of heat equation type and one has solutions for positive time to the initial value problem; in favourable cases it can be shown that the solutions of the evolution equation converge to critical points as time tends to infinity. There is no symmetry between positive and negative time, and – as is well known – we cannot solve the 'backward' heat equation, even for a short time, with general smooth initial data.

The instanton theory we have discussed has quite a different character: the Hessian has infinitely many positive and negative eigenvalues, and this reflects the rough symmetry between positive and negative time (which can be interchanged if we simultaneously reverse the orientation of Y). As with a backward heat equation, we cannot solve the initial value problem for instantons, even for a short time, so the vector field $\mathcal{V}$ does not really define a flow on $\mathcal{B}_P$. Similarly, we do not expect the set of critical points to have any relation to the ordinary homology groups of $\mathcal{B}$.

In this Section we have developed the 3-dimensional point of view in rough parallel with the topics discussed, from the 4-dimensional point of view, in Section 2.2. One topic which we have omitted so far, and which will be developed at length in Chapter 3 below, is the analogue of the Fredholm index theory for the D-operators in four dimensions. We shall see that, roughly speaking, this Fredholm theory goes over to a theory which will allow us to define the *Morse indices* of the critical points of the Chern–Simons function, taking values in $\mathbf{Z}/8$ rather than the usual positive integers. We shall see that this can be regarded as a 'renormalised' count of the (infinite number of) negative eigenvalues, and the ambiguity of 8 will be derived from the factor 8 in the index Formula 2.5. Granted this, we can summarise in the following table a rough dictionary for translating between the 3- and 4-dimensional points of view.

Four dimensions	Three dimensions
X^4	$Y^3 \times \mathbf{R}$
$c_2(P) \in \mathbf{Z}$	$\pi_1(\mathcal{B}_Y) = \mathbf{Z}$
Chern–Weil integrand	Chern–Simons functional θ
$F^+(\mathbf{A}) = \mathbf{0}$	$\frac{\partial A_t}{\partial t} = *_3 F(A_t)$
instantons minimise Yang–Mills functional	$*_3 F$ is the gradient of $\theta : \mathcal{B}_Y \to \mathbf{R}/\mathbf{Z}$
Fredholm index of $D_{\mathbf{A}}$	Morse index of flat connection $\in \mathbf{Z}/8$

2.6 Appendix A: local models

In this Appendix we will explain that the Chern–Simons functional has a finite-dimensional local model, near each critical point (i.e. flat connection). The results are not used in any important way in the book (only in the Appendix to Chapter 4). The discussion is not really specific to the Chern–Simons case. Two general contexts in which the ideas are developed are on the one hand the theory of 'centre manifolds' in PDE theory and on the other hand in the Kodaira–Spencer–Kuranishi deformation theory of geometric structures; the structures in our case being flat connections over 3-manifolds.

We will first fix some notation. To simplify our discussion let us move to a set-up based on fixed Sobolev spaces, for example connections of class L^2_1. We let U be the L^2 orthogonal complement of the harmonic space H^1_A in $T = \ker d^*_A \subset L^2_1$, and U' be the L^2 closure in $T' = \ker d^*_A \subset L^2$. The linear map Q_A from T to T' is the direct sum of an invertible component q mapping U to U' and the zero component on H^1_A. The curvature gives a map $a \mapsto *F(A+a) = *(da + a \wedge a)$ from T to the fixed vector space of L^2 1-forms. However, we really want to consider it as a section of the bundle which assigns the vector space $\ker d^*_{A+a}$ to $a \in T$. To describe it concretely we need a local trivialisation of this bundle, and we obtain this from the L^2 projections

$$\pi_a : \ker d^*_{A+a} \to T = \ker d^*_A.$$

Then the curvature is represented by the map $a \mapsto \pi_a F(A+a)$ from T to T.

Now we have explained in Section 2.1 that a neighbourhood of the point $[A]$ in the quotient space $\mathcal{B}$ is modelled on a quotient of T, dividing by the natural action of Γ_A. We obtain other 'local co-ordinates' by applying diffeomorphisms of T. We say that a diffeomorphism ϕ from a neighbourhood N of 0 in T to itself is 'L^2-compatible' if it distorts the L^2 norm by a bounded amount, i.e. if there is a constant C such that

$$C^{-1}\|\xi\|_{L^2} \leq \|(d\phi)_a\xi\|_{L^2} \leq C\|\xi\|_{L^2},$$

for all a in N. (These are the natural co-ordinate changes in our theory, since we are using the L^2 norm to define the gradient of ϑ.)

Proposition 2.5 (1) *There are an L^2-compatible diffeomeorphism ϕ on a neighbourhood N, a linear isomorphism $l : U \to U'$ and a smooth map $f : N \to H^1_A$, such that for $a = h \oplus b \in N \subset H^1_A \oplus U$,*

$$\pi_{\phi(a)} F(A + \phi(a)) = l(b) + f(h).$$

In particular, if f_0 denotes the restriction of f to H^1_A the diffeomorphism ϕ identifies a neighbourhood of 0 in the zero set of the curvature map $a \to F(A+a)$ in T with a neighbourhood of 0 in the zero set of the finite-dimensional map $f_0 : H^1_A \to H^1_A$.

(2) *In addition we can suppose N and ϕ are chosen so that there is a*

smooth function $\tilde{\vartheta} : H^1_A \to \mathbf{R}$ *and for* $a = h \oplus b \in N$,

$$\vartheta(A + \phi(a)) = \langle b, l(b) \rangle + \tilde{\vartheta}(h).$$

The maps f *and* $\tilde{\vartheta}$ *can be taken to be real analytic, and if* A *is reducible we may choose the maps to commute with the natural actions of* Γ_A.

In brief, this proposition asserts that the Chern–Simons function is locally equivalent to the direct sum of a *non-degenerate quadratic function* and a function on the finite-dimensional space H^1_A.

The idea of these local models will be familiar to readers who have studied the moduli spaces of instantons; see [17][Chapter 4]. The model for the curvature (1) is just like that in the instanton theory; it is obtained by applying the implicit function theorem to the composite of $a \mapsto \pi_a F(A+a)$ with the projection from T' to U'. This is a smooth map with surjective derivative at 0, so it is equivalent under a diffeomorphism of the domain to a linear map, and this gives the required model.

Part (2) is a version of the Morse lemma. We sketch the main steps in the proof. First, suppose that $H^1_A = 0$, so Q_a is an isomorphism and the critical point is non-degenerate in the usual sense. Then we can apply the infinite-dimensional version of the Morse lemma due to Palais, [30]. Since ϑ is a smooth function, on the L^2_1 connections, this result shows that it is locally equivalent to a quadratic function under a diffeomorphism. The only point to check is that this diffeomorphism is L^2-compatible, and this follows readily from an examination of the proof (the derivative of the diffeomorphism differs from the identity by a pseudo-differential operator of order -1). For the general case, when $H^1 \neq 0$, we proceed as follows. The construction of part (1) tells us that after applying a diffeomorphism ϕ (with $d\phi$ equal to the identity at 0) ϑ is equivalent to a function $H^1_A \times U$ such that for each $h \in H^1_A$ the point $(h, 0)$ is a critical point of the restriction of ϑ to the slice $U_h = \{h\} \times U$. When $h = 0$ this is a non-degenerate critical point by construction, so the same is true for all small enough h. So we may apply an obvious parametrised version of the Morse lemma to the family of functions, with h as a parameter, given by restricting ϑ to the U_h. Thus we may assume that ϑ is quadratic + constant on each U_h and, by applying a family of linear maps, that the quadratic part is the same for all h. This then gives the required description as the sum of a function of h and a quadratic form on U.

As a simple example, consider the trivial flat connection A over the 3-torus T^3. Then H^1_A is the tensor product $H^1(T^3) \otimes \mathfrak{su}(2)$. We define a trilinear form on H^1_A by taking the tensor product of the cup product on $H^1(T^3)$ and the form

$$(h_1, h_2, h_3) \mapsto \langle h_1, [h_2, h_3] \rangle$$

on the Lie algebra. The resulting form is symmetric and so defines a cubic function on H^1_A and one can show that, up to a constant, this is the function $\tilde{\vartheta}$ in this case. More generally the same recipe gives the cubic term in the Taylor expansion of $\tilde{\vartheta}$ at the trivial connection on any 3-manifold. If this is degenerate one needs to take account of the higher order terms, which are defined by 'Massey products' on the cohomology.

2.7 Appendix B: pseudo-holomorphic maps

There is a detailed analogy between the differential geometry of

- Yang–Mills connections, and especially Yang–Mills instantons, over a 4-manifold.
- harmonic maps from a Riemann surface to a Riemannian manifold, and especially pseudo-homolomorphic maps to an *almost Kahler manifold.*

This analogy extends to particular phenomena appearing over tubes which we have studied, in the Yang–Mills case, in this Chapter, and there is a version of Floer's theory which applies to the mapping set-up, yielding information about symplectic manifolds. In fact Floer's work in the Yang–Mills case grew out of his work in symplectic geometry (which led most notably to his proof of the 'Arnold conjecture'). In this Appendix we will recall briefly the basic points of this analogy, although we shall not make any use of the ideas in the rest of the book.

We begin with a fixed compact Riemann surface Σ and *symplectic manifold* (M, ω). We define the *degree* of a map $f : \Sigma \to M$ by

$$d = \int_\Sigma f^*(\omega).$$

This is a topological invariant, depending only on the homotopy class of f and the cohomology class of ω. An *almost Kahler* structure on (M, ω) is defined to be an almost complex structure

$$I : TM \to TM, \quad I^2 = -1,$$

such that ω is the imaginary part of a Hermitian metric g on TM. If the almost complex structure is integrable this is precisely a Kahler metric, in the standard sense. These almost Kahler structures exist for any symplectic manifold, and the space of the structures is contractible. Now, just as for any Riemannian target manifold, the *harmonic map* equations for a map from Σ to (M, g) are the Euler–Lagrange equations of the energy functional

$$E(f) = \int_{\Sigma} |df|^2.$$

(Here the notation may be slightly simpler if one chooses a metric within the conformal class on Σ but one can then check that the integral is conformally invariant.) We compare this integral with the one defining the degree of f. At each point of Σ we can decompose the derivative df into complex linear and anti-linear parts

$$df = \partial f + \overline{\partial} f.$$

Then it is easy to see that

$$f^*(\omega) = (|\partial f|^2 - |\overline{\partial} f|^2)\mathrm{vol}_{\Sigma},$$

while $|df|^2 = |\partial f|^2 + |\overline{\partial} f|^2$. So we have

$$d = E^+ - E^-, \quad E = E^+ + E^-,$$

where $E^{\pm}$ are the squares of the L^2 norms of $\partial f, \overline{\partial} f$. We call a map f *pseudo-holomorphic* if it satisfies the analogue of the Cauchy–Riemann equations $\overline{\partial} f = 0$, i.e. if $E^- = 0$. Then we see that *pseudo-holomorphic maps are harmonic, and minimise the energy functional in their homotopy class.* Of course the analogy we are working around is with the Yang–Mills instantons, $F^+(A) = 0$, which minimise the Yang–Mills functional over connections on a given bundle, where the degree of a map is the analogue of the Chern number of a bundle.

We will now go to the case of tubes, so we suppose Σ is a cylinder

$$\Sigma = S^1 \times \mathbf{R},$$

with the standard conformal structure. Of course this is not compact but this is not a serious handicap since one can work with maps which extend to the conformal compactification

$$S^2 = \Sigma \cup \{\pm\infty\}.$$

We can clearly think of a map f from Σ as a one-parameter family γ_t, $t \in \mathbf{R}$, of maps

$$\gamma_t S^1 \to M,$$

i.e. as a path in the infinite-dimensional space LM of loops in M. Just as in the Yang–Mills case, the topological invariant in the 'space-time' picture dimension manifests itself in loop space. For any path γ_t in LM with end points on the constant loops we can associate a map $f : S^2 \to M$ and the degree of f goes over to give a cohomology class

$$\delta \in H^1(LM; \mathbf{Z}).$$

The analogue of the Chern–Simons functional is defined as follows. Suppose for simplicity that M is simply connected and $H_2(M) = \pi_2(M) = \mathbf{Z}$ (for example we could take $M = \mathbf{CP}^n$). We also suppose that ω is an integral cohomology class, so the degrees are all integers. Then any loop γ in M bounds a disc Δ and we define

$$\varphi(\gamma) = \int_\Delta \omega \bmod \mathbf{Z}.$$

The argument to see that this is independent of the choice of discs is the familiar one: gluing two choices along γ to obtain a sphere. Note that φ depends only on the symplectic structure.

Now the space LM has a standard Riemannian metric. A tangent vector to LM at a loop γ is a section ξ of $\gamma^*(TM)$ over S^1 (a vector field 'along γ') and we can take the usual L^2 inner product $\langle\, , \rangle$, using the pull-back of g. We claim that *the gradient flow lines of the* $(\mathbf{R}/\mathbf{Z})$*-valued function* φ *correspond to the pseudo-holomorphic maps of the tube.* To see this we first note that the derivative $(d\varphi)_\gamma$ – a cotangent vector on LM – is the map

$$\xi \mapsto \int_{S^1} \omega(\xi, \gamma').$$

Here γ' is the velocity vector of γ and we have dropped the distinction between TM and its pull-back $\gamma^*(TM)$ in our notation. This formula follows immediately from the definition of φ, if one thinks of extending the disc Δ by a small collar. Now from the algebraic relation between the symplectic, complex and metric structures, we can write the cotangent vector above as

$$\xi \to \langle \xi, I\gamma' \rangle.$$

So the gradient of φ is $I\gamma'$ and a curve γ_t is a gradient line if

$$\frac{d}{dt}(\gamma_T) = I\gamma',$$

and these are precisely the Cauchy–Riemann equations defined by I. It may be clearer to introduce a variable s in the circle, so $\gamma' = \frac{d\gamma}{ds}$ and the equation is

$$\frac{d\gamma}{dt} = I\frac{d\gamma}{ds}.$$

This is all we will say about the symplectic case, except to note one facet of the analogy which may help the reader particularly in the analysis of Chapter 2 of this book. There is a linear differential operator over the cylinder associated to any map, obtained by linearising the pseudo-holomorphic condition. These operators are variants of the Cauchy–Riemann operator. For the basic model we take maps into $\mathbf{C}$, when we get the linear operator

$$\overline{\partial} = \frac{d}{dt} + B,$$

where B is the operator $-i\frac{d}{du}$ over the circle. More generally, one could consider this twisted by a flat connection over the circle. This leads to a simple model for the more complicated 3-dimensional situation considered above. The analysis of the next Chapter can all be applied to operators of this form; on the other hand the proofs in the 3-dimensional case may become more transparent if one works by analogy with the simpler situation. For example the decomposition of the space into two infinite-dimensional subspaces, spanned by positive and negative eigenspaces, is just the decomposition of $L^2(S^1)$ into the *Hardy spaces*

$$\begin{aligned} L^2 &= H^+ \oplus H^-, \\ H^+ &= \left\{\sum a_n e^{in\theta} : n \geq 0\right\}, \\ H^- &= \left\{\sum a_n e^{in\theta} : n < 0\right\}. \end{aligned}$$

2.8 Appendix C: relations with mechanics

We begin by considering, in classical mechanics, the motion of a particle of unit mass on a Riemannian manifold M in a potential V. Thus the motion $q(t)$ is governed by Newton's law:

$$\nabla_t \nabla_t q = -\operatorname{grad} V(q),$$

which are the Euler–Lagrange equations for the action integral

$$\int \left(\tfrac{1}{2}|\nabla q|^2 - V(q)\right) dt.$$

The energy of the motion is $\frac{1}{2}|\nabla q|^2 + V(q)$.

Suppose we have found a function σ on M satisfying the differential equation

$$V = -\tfrac{1}{2}|\operatorname{grad}\sigma|^2 \tag{2.25}$$

Then a gradient path $q(t)$ of σ, i.e. a solution of the *first order* equation

$$\nabla_t q = \operatorname{grad}\sigma,$$

is automatically a solution of Newton's second order equation. To see this, suppose for simplicity that the metric on M is Euclidean, then we have

$$\frac{d^2 q_i}{dt^2} = \frac{d}{dt}\left\{\frac{\partial\sigma}{\partial q^i}\right\} = \sum_j \frac{\partial}{\partial q_j}\left\{\frac{\partial\sigma}{\partial q^i}\right\}\frac{dq_j}{dt},$$

which is

$$\sum_j \frac{\partial^2\sigma}{\partial q_i\partial q_j}\,\frac{\partial\sigma}{\partial q_j}.$$

On the other hand the gradient of V is given by

$$\begin{aligned}\frac{\partial V}{\partial q_i} &= -\frac{\partial}{\partial q_i}\sum_j\left(\frac{\partial\sigma}{\partial q_j}\right)^2\\ &= -2\sum_j \frac{\partial^2\sigma}{\partial q_i\partial q_j}\frac{\partial\sigma}{\partial q_j}.\end{aligned}$$

We can also derive this fact directly from the Lagrangian. For any path $q(t)$, with given end points, we have

$$\int\left(\tfrac{1}{2}|\nabla_t q|^2 - V(q)\right)dt = \int\left(\tfrac{1}{2}|\nabla_t q - \operatorname{grad}\sigma|^2 - \left(\frac{d\sigma}{dt}\right)\right)dt \tag{2.26}$$

$$= \tfrac{1}{2}\int|\nabla_t q - \operatorname{grad}\sigma|^2 dt - \delta\sigma. \tag{2.27}$$

Since the variation $\delta\sigma$ in σ is fixed by the end conditions the Lagrangian is equivalent to that given by the first integral – the integral of $|\nabla_t q - \operatorname{grad}\sigma|^2$ – which is plainly an extremum, and indeed an absolute minimum, for the gradient path.

The solutions of the equation of motion found in the manner above are characterised by the fact that their total energy $\frac{1}{2}|\nabla_t q|^2 + V$ is zero. The potential V is not unique, we can change it by a constant without changing the mechanical problem. Thus we can immediately generalise Equation 2.25 to a family of equations:

$$\tfrac{1}{2}|\operatorname{grad}\sigma|^2 + V = E, \tag{2.28}$$

with a parameter E.

The geometric meaning of this construction is clarified by considering the particular case of free motion on M, with $V = 0$. Then if we fix a reference point p in M we can obtain solutions $\sigma_E(q) = \frac{1}{\sqrt{E}} d(q, p_0)$ from the Riemannian distance function of M (at least in a neighbourhood of p_0). In general, when the potential is present, for a given energy $E \geq 0$, and for points q close to p_0, we can suppose there will be a unique particle motion with energy E which starts at p_0 at time 0 and passes through q at some subsequent time $\tau(q, E)$. Then if we define $\sigma_E(q)$ to be the action of this path we obtain solutions of Equation 2.28 whose corresponding gradient lines are the motions through p_0.

Equation 2.28 is closely related to the *Hamilton–Jacobi equation* for a function $S(q, t)$:

$$\frac{\partial S}{\partial t} = \tfrac{1}{2}|\nabla S|^2 + V. \tag{2.29}$$

In the framework above, a solution $S(q, t)$ is obtained as the action of the motion beginning at p_0 at time 0, and arriving at q at time t.

We will now explain the relation between Yang–Mills theory and this classical mechanical picture. We consider the Yang–Mills equations over 'space-time' $Y \times \mathbf{R}$, but first with respect to the Lorentzian metric $dy^2 - dt^2$. We decompose the curvature F of a connection over the tube into 'electric' and 'magnetic' components E, B:

$$F = *B + E \wedge dt.$$

The shift to the Lorentzian metric means that the Lagrangian for the Yang–Mills equation is the indefinite expression

$$\int_{Y \times \mathbf{R}} (|E|^2 - |B|^2),$$

and the Lorentzian Yang–Mills equations themselves can be written

$$d^{*,L}_{\mathbf{A}} F_{\mathbf{A}} = 0,$$

where $d^{*,L}$ is the formal adjoint defined by the Lorentzian metric. A moment's thought shows that the component $i_t d^{*,L}$ is the same as $i_t d^*$, formed from the Euclidean metric. Thus Yang–Mills solutions, in either signature, yield horizontal curves of the kind considered in Section 2.3.

So, as we explained above, we can identify the solutions with a subset of the paths $[A_t]$ in $\mathcal{B}_P$. Now the magnetic component B of the curvature is just the curvature of the connection A_t restricted to Y while, as we have seen, the electric term can be identified with the *velocity vector* of the path in $\mathcal{B}_P$. So the Lorentzian Yang–Mills Lagrangian can be written, in terms of paths in $\mathcal{B}_P$, as

$$\int \left(\|\nabla_t A_t\|^2 - V(A_t)\right) dt,$$

where the potential function V on $\mathcal{B}_P$ is the 3-dimensional Yang–Mills action

$$V(A) = \int_Y |F_A|^2.$$

We see then that the 4-dimensional Lorentzian Yang–Mills solutions can be regarded as *the motions of a particle moving on the Riemannian manifold $\mathcal{B}_P$ in the potential $V = \|F\|^2$*. As a simple model to have in mind, in the case when Y is the 3-sphere we can picture V as a well, with a minimum at the trivial flat connection, and we can think of Lorentzian Yang–Mills solutions as representing oscillations about this minimum. To obtain the Euclidean Yang–Mills solutions in this framework we merely have to change the sign of the potential, so that for example the trivial flat solution now becomes highly unstable. (This reflects the fact that the initial value problem is ill-posed in the Euclidean case.)

We can now fit the discussion of Section 2.3 into the general framework in mechanics reviewed above. Locally in $\mathcal{B}_P$ we can regard the Chern–Simons function ϑ as a real-valued function and we have

$$|\operatorname{grad}\vartheta|^2 = \|F\|^2 = V.$$

So the Chern–Simons function $\sigma = \vartheta$ satisfies Equation 2.25 in this infinite-dimensional setting, but with the *reversed potential*, and the fact that the gradient lines (i.e. the Yang–Mills instantons) give solutions to the Euclidean Yang–Mills equations becomes a special case of the general principle above. Moreover, the fact that the instantons are absolute minima of the Euclidean action, with suitable end conditions, becomes a special case of Equation 2.26.

The utility of this point of view may seem to be limited by the fact that we have to reverse the potential on $\mathcal{B}$ from the physical, Lorentzian, case. To give physical motivation to this we have to pass beyond classical mechanics to a quantum mechanical picture. Thus we return to our general model on the manifold M and consider the Schrödinger equation:

$$h^2\Delta\varphi + V\varphi = E\varphi, \tag{2.30}$$

where h (Planck's constant) is a small parameter. The solutions are the wave functions which give the energy eigenstates of the quantum mechanical version of the motion of particle in the potential V. We want to recall the main ideas of the *short wave* or *quasi-classical* or *WKJB* asymptotic description of solutions of the Schrödinger equation, for small values of h. We begin by considering the region in M where $V < E$, i.e. the region accessible by particles with energy E governed by the classical equation. We seek 'wavelike' solutions in the form

$$\phi(x) = A(x)e^{ih\sigma(x)},$$

where σ is real. If we substitute this into Equation 2.30 we get, after a short calculation,

$$h^{-2}(|\nabla\sigma|^2 + (V - E))A + h^{-1}(\nabla\cdot(A\nabla\phi) + 2(\nabla A)\cdot(\nabla\phi)) + \Delta A = 0 \tag{2.31}$$

The idea is that for small h we get good approximations to the solutions by imposing the leading terms of Equation 2.31 separately, i.e. by looking first at solutions σ of

$$|\nabla\sigma|^2 = E - V,$$

which is just Equation 2.28 that we have studied in the classical setting. We then look at amplitude functions A which make the h^{-1} term vanish,

$$\nabla\cdot(A\nabla\phi) + 2(\nabla A)\cdot(\nabla\phi) = 0.$$

If σ is given this becomes an ordinary differential equation

$$\nabla_v A = -3(\Delta\sigma)A$$

for A along the trajectory of the vector field $v = \operatorname{grad}\sigma$, which, as we have seen, corresponds to a classical motion with energy E. The basic idea then is that we can approximate the quantum wave functions for the problem by integrating this ODE along a suitable family of classical trajectories. More generally we can consider linear combinations of these functions, with different choices of σ and A. There is a similar discussion relating the time-dependent Schrödinger equation

$$h^2\nabla\phi + V\phi = i\frac{d\phi}{dt},$$

to the Hamilton–Jacobi Equation 2.29.

We now move to the region, inaccessible in the classical picture, where $V > E$. The appropriate asymptotic model takes the form

$$\phi = Ae^{h\sigma},$$

and in the same fashion as before we obtain the equations

$$|\nabla\sigma|^2 = (V - E), \quad \nabla \cdot (A\nabla\phi) + 2(\nabla A) \cdot (\nabla\phi) = 0.$$

This gives us asymptotic model solutions by integrating an ODE along a suitable family of gradient lines of a solution σ to the first equation; i.e. along a suitable family of *classical trajectories of particle motions with energy* $-E$ *in the reversed potential* $-V$. Rather than the oscillatory, wave-like, solutions in the first case the solutions are now typically exponentially growing or decaying along the gradient lines, and the relevant solutions are picked out by the details of the problem. For example, consider the simple 1-dimensional case of motion in a potential well. To get an L^2 wave function we have to pick out the decaying solutions in each component of the region $V > E$. The amplitudes of these, and the relevant wave-like solution in the region $V < E$, are determined by a more sophisticated analysis of 'matching conditions' in the transition regions between the two regimes (cf. [40] in the 1-dimensional case).

The idea we hope to have brought out in this brief sketch is that the solutions of the classical problem with the reversed potential are quite relevant in quantum theory; and at least by analogy we expect the classical solutions of the Riemannian Yang–Mills equations to be relevant to the *quantum field theory*, i.e. quantum mechanics on $\mathcal{B}_Y$. We will refer the reader to proper references for more details; but we should note that a particularly important application occurs in the quasi-classical modelling of *tunnelling* in quantum mechanics. For this we consider a case when the potential V has two minima, say at q_0 and q_1 where $V(q_0) = V(q_1) = 0$, separated by an energy barrier of level $E_0 > 0$. The classical motions with energy $E < E_0$ fall into two classes, located near either q_0 or q_1, but in the quantum picture a particle located at one time near q_0 can tunnel through the barrier to be found at q_1 at a later time, even if the energy is less than E_0. Another way of expressing this is that the eigenspace belonging to the least energy eigenstate E is 1-dimensional although there is a second eigenvalue $E' \sim E + Ce^{-D/h}$ near by. The gap between the eigenvalues measures the strength of the tunnelling, and the co-efficients C, D in the asymptotic approximation for the gap can be found from the *instantons* for the quantum mechanical problem. By definition these are the solutions $q(t)$ of the classical equations for a particle moving in potential $-V$ with energy 0, with

$$q(t) \to q_0 \text{ as } t \to -\infty, \quad q(t) \to q_1 \text{ as } t \to \infty.$$

In our infinite-dimensional case q_0, q_1 become flat connections and the 'tunnelling instantons' with minimal action are just the Yang–Mills instantons over the tube with these flat limits. Connections of this kind are in turn precisely the central object of study in Floer's theory.

3
Linear analysis

We will now begin work in earnest, and develop the basic analytical material which underpins Floer's theory. This consists principally of results on linear differential operators over manifolds with tubular ends, results which go back at least to the work of Atiyah, Patodi and Singer [4] in the early 1970s. Another general reference is [33]. These manifolds are not compact, so we cannot appeal to the familiar package of theorems for elliptic operators over compact manifolds directly. Although analysis on general non-compact manifolds can present many serious difficulties the manifolds we have to deal with are, comparatively, very simple. Standard techniques bring the main questions immediately down to the case of operators over the tubes themselves. Here one can use separation of variables to bring the usual results of elliptic theory over the compact cross-section of the tube to bear. The special 'non-compact' features are effectively reduced to problems of *ordinary* differential equations over the real line.

3.1 Separation of variables

Let us begin then by supposing that, as in the previous Chapter, A is a flat $SU(2)$ connection over a compact 3-manifold Y, that $\mathbf{A}$ is the corresponding connection on the pull-back bundle $\mathbf{P} = \pi^*(P)$ over $Y \otimes \mathbf{R}$ and that $D = D_{\mathbf{A}} = d^*_{\mathbf{A}} + d^+_{\mathbf{A}}$ is the deformation operator of $\mathbf{A}$ (acting on the forms with values in the associated adjoint bundle $\mathfrak{g}_P$). We have seen that, regarded as a map from the space of sections of the pull-back of $(\Lambda^0_Y \oplus \Lambda^1_Y)(\mathfrak{g}_P)$ to itself, this operator can be written as

$$D = \frac{d}{dt} + L,$$

where $L = L_A$ is the differential operator $d_A^* + d_A$ over Y. We shall assume that A is an acyclic connection, so that L is an invertible operator; there is thus a complete eigenspace decomposition ϕ_λ, with $L\phi_\lambda = \lambda\phi_\lambda$, where λ may have either sign but $|\lambda| \geq \delta$ for some $\delta > 0$.

We use separation of variables to prove two key lemmas.

Lemma 3.1 *Let ρ be a smooth, compactly supported, section of $(\Lambda^0 \oplus \Lambda^+)(\mathfrak{g}_P)$ over $Y \times \mathbf{R}$. There is a smooth section f of $(\Lambda^0 \oplus \Lambda^1)(\mathfrak{g}_P)$ which satisfies the equation $Df = \rho$, and with $\|f\| \leq \delta^{-1}\|\rho\|$.*

To prove this we write

$$\rho = \sum_\lambda \rho_\lambda(t)\phi_\lambda,$$

where ρ_λ are smooth, compactly supported, functions on $\mathbf{R}$. Then we seek a solution $f = \sum f_\lambda(t)\phi_\lambda$.The equation becomes

$$Df = \left(\frac{d}{dt} + L\right)\left(\sum f_\lambda\phi_\lambda\right) = \sum\left(\frac{d}{dt}f_\lambda + \lambda f_\lambda\right)\phi_\lambda = \sum \rho_\lambda\phi_\lambda.$$

This is satisfied if

$$\frac{d}{dt}f_\lambda + \lambda f_\lambda = \rho_\lambda$$

for each λ. But it is elementary that this latter equation has a bounded solution, for $\lambda \neq 0$. If $\lambda < 0$ the solution is

$$f_\lambda(t) = e^{-\lambda t}\int_{-\infty}^t e^{\lambda\tau}\rho_\lambda(\tau)\,d\tau,$$

and if $\lambda < 0$ it is

$$f_\lambda(t) = -e^{\lambda t}\int_t^\infty e^{-\lambda\tau}\rho_\lambda(\tau)\,d\tau.$$

In either case the solution decays as $e^{-|\lambda t|}$ at infinity (and in fact, vanishes on one half-line).

We have also

$$\begin{aligned}\left(\frac{d}{dt}f_\lambda\right)^2 + \lambda^2 f_\lambda^2 &= \left(\frac{d}{dt}f_\lambda + \lambda f_\lambda\right)^2 - \lambda\frac{d}{dt}(f_\lambda)^2 \\ &= \rho_\lambda^2 + \frac{d}{dt}(f_\lambda^2).\end{aligned}$$

Since f_λ decays at infinity we can integrate this over the line to obtain

$$\int_{-\infty}^\infty \left(\frac{d}{dt}f_\lambda\right)^2 + \lambda^2 f_\lambda^2\,dt = \int \rho_\lambda^2\,dt.$$

In particular

$$\|f_\lambda\|^2_{L^2(\mathbf{R})} \leq \delta^{-2}\|\rho_\lambda\|^2_{L^2(\mathbf{R})}.$$

Hence the sum $\sum_\lambda f_\lambda \phi_\lambda$ defines an L^2 section f of $(\Lambda^0 \oplus \Lambda^1)(\mathfrak{g}_P)$ over $Y \times \mathbf{R}$, with

$$\|f\|^2_{L^2(Y\times\mathbf{R})} = \sum_\lambda \|f_\lambda\|^2_{L^2(\mathbf{R})} \leq \delta^{-2}\sum \|\rho_\lambda\|_{L^2(\mathbf{R})} = \|\rho\|^2_{L^2(Y\times\mathbf{R})}.$$

It is quite routine to show that the proposed solution f does indeed satisfy the equation $Df = \rho$, and is smooth. Indeed, it follows easily from the construction that f is a weak solution of the equation, and then we can use elliptic regularity to verify smoothness. Thus the proof of the lemma is complete.

The separation of variables also gives more or less immediately that f is the *unique* L^2 solution to the equation, i.e. the only L^2 solution to the equation $Df = 0$ is $f = 0$. However, we will deduce this from a rather stronger, quantitative, result. For $T > 1$ we consider the finite tube $Y \times (-T, T)$ and two 'bands':

$$B_T^+ = Y \times (T-1, T),\ B_T^- = Y \times (-T, -T+1).$$

Lemma 3.2 *Let f be a solution of $Df = 0$ over the finite tube $Y \times (-T, T)$. Then*

$$\int_{Y\times(-T,T)} |f|^2 \leq c_\delta \left(\int_{B_T^+} |f|^2 + \int_{B_T^-} |f|^2 \right),$$

where $c_\delta = \frac{1}{1-e^{-2\delta}}$.

For the proof we again expand f as $f = \sum f_\lambda \phi_\lambda$, with functions f_λ on $(-T, T)$ which satisfy $\frac{d}{dt}f_\lambda + \lambda f_\lambda = 0$. Hence $f_\lambda(t)$ can be written $f_\lambda(t) = a_\lambda e^{-\lambda t}$, for a constant a_λ. The point will be that the integral over the band B_T^+ controls the f_λ for $\lambda < 0$ and the integral over the other band controls those for $\lambda > 0$. Consider first the eigenvalues $\lambda > 0$, for which f_λ decay for positive t. We have

$$\int_{-T}^{T} |f_\lambda|^2\, dt = a_\lambda^2 \int_{-T}^{T} e^{-2\lambda t}\, dt < a_\lambda^2 \int_{-T}^{\infty} e^{-2\lambda t}\, dt = a_\lambda^2 \frac{e^{2\lambda T}}{\lambda}.$$

Similarly $\int_{-T}^{-T+1} |f_\lambda|^2\, dt = a_\lambda^2 \frac{e^{2\lambda T} - e^{2\lambda(T-1)}}{\lambda}$, so

$$\int_{-T}^{T} |f_\lambda|^2\, dt \leq \frac{1}{1-e^{-2\lambda}} \int_{-T}^{-T+1} f_\lambda^2 \leq \frac{1}{1-e^{-2\delta}} \int_{-T}^{-T+1} f_\lambda^2.$$

We now write $f = f_+ + f_-$, where

$$f_+ = \sum_{\lambda>0} f_\lambda \phi_\lambda, \quad f_- = \sum_{\lambda<0} f_\lambda \phi_\lambda.$$

Summing the previous inequality over the positive spectrum we get

$$\int_{Y\times(-T,T)} |f_+|^2 \leq c_\delta \sum_{\lambda>0} \int_{-T}^{-T+1} f_\lambda^2 \, dt = c_\delta \int_{B_T^-} |f_+|^2.$$

Similarly, for the negative spectrum, we have

$$\int_{Y\times(-T,T)} |f_-|^2 \leq \frac{1}{e^{2\delta}-1} \int_{B_T^+} |f_-|^2.$$

Hence

$$\begin{aligned} \int_{Y\times(-T,T)} |f|^2 &= \int_{Y\times(-T,T)} (|f_+|^2 + |f_-|^2) \\ &\leq \frac{1}{1-e^{-2\delta}} \left(\int_{B_T^+} |f|^2 + \int_{B_T^-} |f|^2 \right) \end{aligned}$$

(since $e^{2\delta} - 1 > 1 - e^{-2\delta}$), as required.

Note that Lemma 3.2 immediately gives that an L^2 solution f to $Df = 0$ defined over the infinite tube $Y \times \mathbf{R}$ is zero, since the integral of $|f|^2$ over the bands B_T^+, B_T^- tends to zero as T tends to infinity, and the constant c_δ is independent of T.

Simple extensions of the proof of Lemma 3.2 give pointwise bounds on all derivatives of the solution, decaying *exponentially* with T.

Lemma 3.3 *For any integer $l > 0$ and positive $s < 1$ there is a constant C such that for all $T > 1$ and solution f over the finite tube, as in Lemma 3.2,*

$$|\nabla^{(l)} f_{y,t}| \leq C e^{-\delta(T-|t|)} \left(\int_{B_T^+} |f|^2 + \int_{B_T^-} |f|^2 \right)$$

for all $|t| \leq T - s$.

To prove this we observe first that by the usual elliptic estimates, applied to a fixed band, it suffices to prove that for $t \leq T - s$

$$\int_{Y\times(t-s,t+s)} |f|^2 \leq 2e^{-\delta(T-|t|)} \left(\int_{B_T^+} |f|^2 + \int_{B_T^-} |f|^2 \right),$$

and this follows easily from the corresponding estimates for the f_λ.

One can obtain from Lemmas 3.1 and 3.2 corresponding results for solutions over a half-line – just consider a sequence of finite tubes – and we conclude that any L^2 solution over a half-line must decay exponentially. In particular, this applies to the solutions of Lemma 3.1, outside the support of ρ. Similarly, the proof of Lemma 3.2 extends easily to show that if $Df = \rho$ over $Y \times (-T, T)$ then

$$\|f\|_{L^2}^2 \leq \delta^{-2}\|\rho\|^2 + c_\delta(\|f|_{B_T^+}^2 + \|f|_{B_T^-}\|^2).$$

It is probably not necessary to point out that in the proofs above we have only used very formal properties of the situation. Lemma 3.1 extends to a general operator over a tube $Y \times \mathbf{R}$ of the form $\frac{d}{dt} + L$ where L is an elliptic operator over Y. In the proofs of Lemmas 3.1 and 3.2 we encounter the decomposition into the spans of the positive and negative eigenspaces of L, which occurs for any self-adjoint operator and in particular for 'Dirac type' operators over odd-dimensional manifolds Y. The paradigm is the operator $\frac{d}{du}$ over the circle, which we met in the discussion of Appendix B to the previous Chapter. Then we get the decomposition into 'Hardy spaces',

$$L^2 = H^+ \oplus H^-,$$

of functions with only positive or negative terms in their Fourier series. (Of course in this case there is a zero eigenvalue, so we have to make a suitable convention for the constants.) In the general case we get a decomposition over Y

$$f = f_+ + f_-,$$

where f_+ extends to an L^2 (and in fact exponentially decaying) solution $Df_+ = 0$ over $Y \times \mathbf{R}^+$ and f^- to a decaying solution over $Y \times \mathbf{R}^-$. In the paradigm above this becomes the Hardy space decomposition of a function on the equator in the Riemann sphere into the boundary values of holomorphic functions over the upper and lower hemispheres.

We now define a Hilbert space L^2_1 to be the completion of the smooth, compactly supported, $\mathfrak{g}_P$-valued 1-forms over the cylinder in the norm

$$\|f\|_{L^2_1}^2 = \|Df\|^2 + \|f\|^2,$$

where, as usual in this book, a norm symbol without further specification denotes an L^2 norm. (We also simplify our notation by omitting explicit reference to the bundle whose sections we are considering.) The differential operator D trivially extends to give a bounded map

$$D : L^2_1 \to L^2,$$

and the content of Lemma 3.1 is the assertion that this is invertible. Thus we have

Proposition 3.4 *The operator $D : L^2_1 \to L^2$ is an isomorphism: there is an inverse $Q : L^2 \to L^2_1$ with*

$$\|Q\rho\|_{L^2} \leq \delta^{-1}\|\rho\|_{L^2}$$

for $\rho \in L^2$.

We define Q on the smooth, compactly supported sections by the formulae of Lemma 3.1. The uniform bound on the L^2 norm of the solution constructed there means that Q extends, by continuity, to the completions and $DQ = 1$. The operator is a two-sided inverse since the kernel of D in L^2_1 is zero: any L^2 solution of $Df = 0$ is smooth, by elliptic regularity for the operator D, and hence is identically zero by the discussion following Lemma 3.2.

3.1.1 Sobolev spaces on tubes

There are a number of definitions of the 'Sobolev space' L^2_1 which turn out to be equivalent. Since it is often useful to apply these different definitions we shall review them now.

First, we could have equivalently defined L^2_1 as the space of L^2 bundle-valued 1-forms f over $Y \times (-\infty, \infty)$ such that Df, defined distributionally, is also in L^2. The equivalence of these points of view is part of the standard equivalence of 'weak and strong' definitions of derivatives. It is obvious that the space defined by the first ('strong') definition is contained in that defined by the second; what needs to be shown is that any L^2 section f with Df in L^2 can be approximated in the given norm by smooth, compactly supported, sections. The proof of this requires two steps. In the first step we approximate any compactly supported f by a sequence of smooth sections. This is achieved by a convolution and is already part of the standard theory over compact manifolds. The other step, in which we approximate a general element by compactly supported ones, is more relevant to us here. The corresponding statement is valid for any *first order operator* over a *complete* Riemannian manifold, where the symbol of the operator is bounded uniformly relative to the metric, in the obvious sense. We introduce smooth functions β_T which are supported in the compact set $[-(T+1), T+1] \subset \mathbf{R}$, equal to 1 over $[-(T-1), (T-1)]$, and with $|\nabla \beta_T| \leq 1$ everywhere. Thus β_T is a smoothing of the characteristic function of $Y \times [-T, T]$.

Then

$$D(\beta_T f) = \beta_T Df + (\nabla \beta_T) * f,$$

where $*$ denotes an algebraic operation. The first term on the right hand side converges to Df in L^2 and the second tends to 0, since its norm is bounded by the integral of $|f|^2$ over the two bands supporting $\nabla \beta_T$, which tends to zero with T if f is in L^2.

Another equivalent definition of the space L^2_1 uses the full covariant derivative ∇, defined by the connection $\mathbf{A}$, in place of the differential operator D. We have a Weitzenbock formula:

$$D^* Df = \nabla^* \nabla f + Rf,$$

where R is a curvature term, which is uniformly bounded over $Y \times \mathbf{R}$. Thus for smooth compactly supported sections f

$$\int_{Y \times \mathbf{R}} |Df|^2 = \int_{Y \times \mathbf{R}} |\nabla f|^2 + (Rf, f).$$

It follows that the norm given by the square root of

$$\|\nabla f\|^2 + \|f\|^2$$

is equivalent to the L^2_1 norm defined above, on the compactly supported sections, and so the two norms define the same completion L^2_1. Of course the argument of the first part shows that we can get the same space by taking the L^2 sections f such that ∇f is in L^2.

Another application of these ideas is to the formal adjoint operator. We can define the formal adjoint operator D^* as a map,

$$D^* : L^2_1 \to L^2.$$

There is a complete symmetry between D and D^*; indeed, since L is self-adjoint on Y we have

$$D = \frac{d}{dt} + L, \quad D^* = -\frac{d}{dt} + L$$

so the operators are interchanged by switching the co-ordinate t to $-t$ (which changes the orientation on the tube). The fact that we shall use often is that if f, g, Df and D^*g are all in L^2 the integration by parts formula

$$\langle Df, g\rangle = \langle f, D^*g\rangle \tag{3.1}$$

is valid. This follows immediately from the fact established above that the compactly supported sections are dense in L^2_1. Expressed more directly, we apply the formula to the functions $\beta_T f$ and g and estimate the error term as above. (We can weaken the hypothesis for Equation 3.1 to hold. All we need is that the pointwise inner product (f, g) is in L^1 over the tube, so that

$$\int_{Y\times\mathbf{R}} (\nabla\beta_T * f, g) \to 0$$

as $T \to \infty$. Then Equation 3.1 holds in the sense that if the integral defining one term converges absolutely then so does that defining the other, and the two integrals are equal.)

3.2 The index

We will now develop an index theory for operators over 4-manifolds with tubular ends. Thus we wish to associate an 'analytical index' to suitable 'topological data'. We begin with a convenient definition. Let X be a 4-manifold with tubular ends, as defined in Chapter 2.

Definition 3.5 *An* adapted *bundle* $\mathbf{P}$ *over* X *is a smooth bundle with a fixed flat connection over each end. Two adapted bundles are* equivalent *if there is a bundle isomorphism between them which preserves the flat structures over the ends.*

Obviously we have a similar notion for a manifold with boundary. To any adapted bundle and end $Y \times \mathbf{R}^+$ of X we can associate a limit connection – a flat connection A over Y – and we can naturally identify the restriction of $\mathbf{P}$ to the end with the pull-back of a bundle P over Y. Notice that, while any $SU(2)$ bundle over X is trivial as a C^∞ bundle, this is not true of adapted bundles, even with the same limiting flat connections. For example, it is easy to see that an adapted bundle over X with all limits trivial is equivalent to an ordinary bundle $\hat{\mathbf{P}}$ over the compactified space $\hat{X}$ obtained by adding a point at infinity to each end to make a cone singularity. Thus we have an extra invariant given by the Chern number

$$\langle c_2(\hat{\mathbf{P}}), [\hat{X}]\rangle,$$

where $[\hat{X}]$ is the natural fundamental class of $\hat{X}$. More generally it is easy to see that the equivalence classes of adapted bundles with given flat limits are put into one-to-one correspondence with the integers by a relative Chern number. This can be defined by choosing an adapted connection on $\mathbf{P}$ – i.e. a connection $\mathbf{A}$ which agrees with the given flat structures over the ends. Then the integral

$$\frac{1}{8\pi^2}\int_X \mathrm{Tr}(F_{\mathbf{A}}^2)$$

is an invariant of $\mathbf{P}$ only (which reduces mod $\mathbf{Z}$ to the sum of the Chern–Simons invariants of the limits).

With these preliminaries in place we proceed to set up our index problem. Let $\mathbf{P}$ be an adapted bundle X with an adapted connection $\mathbf{A}$ as above. For the moment we assume that *each of the limiting flat connections is acyclic.* We define the Sobolev space L_1^2 to be the completion of the smooth compactly supported forms under the norm

$$\left(\|\nabla_{\mathbf{A}} f\|_{L^2}^2 + \|f\|_{L^2}^2]\right)^{1/2}.$$

The deformation operator $D = D_{\mathbf{A}}$ extends to a Hilbert space completion to give a bounded operator $D : L_1^2 \to L^2$ just as before (and the same remarks on the various different definitions of the space L_1^2 hold good). We also have a formal adjoint operator D^*, and the integration-by-parts Equation 3.1 is valid for sections of the appropriate kind. Notice that the domains of D and D^* are quite distinct in general. In a more detailed notation:

$$D^* : L_1^2(\Omega_X^0 \oplus \Omega_X^+)(\mathfrak{g}_P)) \to L^2(\Omega_X^1(\mathfrak{g}_P)),$$

$$D : L_1^2(\Omega_X^1)(\mathfrak{g}_P) \to L^2((\Omega_X^0 \oplus \Omega_X^+)(\mathfrak{g}_P)).$$

(We shall not usually be so explicit in our choice of notation.)

Now it need not be true that the operator D is invertible, as happened in the case of tubes themselves. Instead, as we shall see next, the tube theory shows that D is a *Fredholm operator.* This is just what we would have, as a matter of routine, over a compact base manifold. For compact manifolds one obtains this Fredholm property by piecing together a finite number of inverses defined in local charts, in which the operator is modelled on a constant co-efficient operator over Euclidean space. In our case the proof is much the same except that we have two kinds of models – the familiar Euclidean ones, in a system of charts covering a compact interior portion of the manifold, and models over the ends in which we use the tube theory.

Proposition 3.6 *For any adapted connection* $\mathbf{A}$ *over a 4-manifold* X *with tubular ends, as above, which has acyclic limits over each end, the operator* $D : L^2_1 \to L^2$ *is Fredholm. That is*

(i) $\ker D \subset L^2_1 = \ker D \cap L^2$ *is finite-dimensional,*
(ii) *the image of* D *is a closed subspace of finite codimension in* L^2*.*

Moreover, the cokernel $L^2/\operatorname{Im} D$ *is isomorphic to the kernel of* D^* *in* L^2_1*.*

This is the extension of the basic global theorem for elliptic operators from the compact case to manifolds with tubular ends. For the proof we can choose from a number of standard approaches, all of which involve much the same ingredients.

To prove (i) we consider first the operator D over a half-tube $Y \times (0, \infty)$, of the kind occurring as an end of X. Let B be the band $Y \times (0, 1)$ and recall that Lemma 3.2 gives an inequality

$$\int_{Y \times (0,\infty)} |f|^2 \leq c_\delta \int_B |f|^2$$

for any solution f of $Df = 0$ over the half-tube. In particular the L^2 norm of a solution f of $Df = 0$ over all of X is controlled by that over a fixed compact subset X_0 of X.

Now we show that the kernel of D on $\mathbf{P}$ over X has finite dimension by showing that the unit ball, in the L^2 norm, of the kernel is compact. Let f_i be a sequence of sections over X with $\|f_i\|_{L^2} \leq 1$ and $Df_i = 0$. By the standard theory for the elliptic operator D over compact sets we may (taking a subsequence) suppose that the f_i converge in L^2 over compact subsets to a limit f_∞. The limit is in L^2, with $\|f_\infty\|_{L^2} \leq 1$, and satisfies $Df_\infty = 0$. If we apply Lemma 3.2 to the differences $f_i - f_\infty$ we see that f_i converges in L^2 over each tubular end and hence over X as required.

We now turn to the proof of the second assertion of Proposition 3.6, using the 'parametrix' method. Again, let $Y \times (0, \infty)$ be an end of X and let Q be the inverse operator given by Proposition 3.4. For any section ρ over X we define $\hat{Q}(\rho)$ by restricting ρ to $Y \times (0, \infty) \subset X$, applying Q, and restricting again to $Y \times (0, \infty) \subset Y \times \mathbf{R}$. Thus $\hat{Q}$ is a bounded operator, with respect to L^2 norms, and

$$D\hat{Q}(\rho) = \rho \quad \text{on } Y \times (0, \infty).$$

Now cover X_0 by a finite number of small Euclidean patches. Appealing to standard, local, elliptic theory we can choose these so that D has a right inverse over each patch; so combining with the inverses over the finite number of ends of X we arrive at a situation where we have a cover $X = \bigcup_{i=1}^n U_i$ such that each intersection $U_i \cap U_j$ is pre-compact in X, and L^2-bounded operators $\hat{Q}_i$ such that $D\hat{Q}_i\rho = 0$ on a neighbourhood of the closure of U_i. Let β_i be a partition of unity subordinate to this cover and define an operator $P : L^2 \to L^2$ by

$$P(\rho) = \sum_{i=1}^{n} \beta_i \hat{Q}_i(\rho).$$

Note that $P(\rho)$ is smooth, if ρ is, unlike $\hat{Q}_i(\rho)$. Also, $DP(\rho) - \rho$ is supported on the overlaps $U_i \cap U_j$, for $i \neq j$. On a given element, U_1 say, of the cover we write

$$P(\rho) = \hat{Q}_1(\rho) + (\beta_1 - 1)\hat{Q}_1(\rho) + \sum_{i=2}^{n} \beta_i \hat{Q}_i(\rho) = \hat{Q}_1(\rho) + \sum_{i=2}^{n} \beta_i (\hat{Q}_i - \hat{Q}_1)(\rho),$$

using the fact that $\sum \beta_i = 1$. Thus, on U_1, we have

$$D\ P(\rho) = \rho + D\left(\sum \beta_i f_i\right),$$

where $f_i = (\hat{Q}_i - \hat{Q}_1)(\rho)$, so $Df_i = 0$. Note that, by the standard local elliptic estimates, all derivatives of f_i over the support of β_i are bounded by fixed multiples of the L^2 norm of f_i, and hence by multiples of the L^2 norm of ρ, since $\hat{Q}_1$ and $\hat{Q}_i$ are bounded in L^2 operator norm. So if we write

$$DP(\rho) - \rho = S(\rho),$$

all derivatives of $S(\rho)$ over U_1 are bounded by multiples of $\|\rho\|_{L^2}$. Clearly the same is true for all the other elements of the cover. Since $S(\rho)$ is supported in a fixed compact set the Ascoli–Arzelà theorem implies that S is a *compact* operator from L^2 to L^2. A standard result of functional analysis asserts then that the image of $1 + S$ is closed and of finite codimension in L^2. It follows trivially from the factorisation $1 + S = DP$ that the same is true of S, so the proof of (ii) is complete.

Finally, since $\operatorname{Im} D$ is closed the cokernel $L^2/\operatorname{Im} D$ is isomorphic to the L^2-orthogonal complement $\ker D^\perp$. If g is orthogonal to the image of

D then it follows immediately from the definition of the formal adjoint that $D^*(g) = 0$. Conversely, if g is in L^2 and $D^*g = 0$ then

$$\langle Df, g \rangle = \langle f, D^*g \rangle = 0,$$

if $f, Df \in L^2$, by Equation 3.1. Thus g is in the orthogonal complement of the image of D and we have shown that $(\operatorname{Im} D)^\perp = \ker D^* \cap L^2$.

One should notice here one difference between our set-up and the usual compact case: the failure of a 'Rellich lemma' in our function spaces. That is, the embedding $L^2_1 \to L^2$ is *not* compact. To see this one need only consider the translates of a compactly supported section over the tube. Of course, in the usual case this Rellich lemma gives immediately the finite dimensionality of $\ker D$.

We now associate to the adapted bundle with connection the usual Fredholm index of the operator $D_{\mathbf{A}}$

$$\operatorname{ind}(\mathbf{P}) = \dim \ker D - \dim \operatorname{coker} D = \dim \ker D - \dim \ker D^*.$$

This depends at the outset on the particular connection $\mathbf{A}$ chosen, and on the metric on X. However, standard elliptic theory tells us that this index is unchanged by continuous variations in the data through Fredholm operators, and this immediately shows that the index depends only on the adapted bundle $\mathbf{P}$ over the oriented 4-manifold X. We handle variations in the connection and of the metric over compact sets by an argument just like that in the case of closed manifolds. Then we handle variations in the metric on Y by extending Section 2.1 in an obvious way to families of operators $\frac{d}{dt} + L_z$, depending on a real parameter z. If L_z is always invertible these give continuous families of invertible operators over the tube.

3.2.1 Remarks on other operators

We have now achieved our main goal of associating an analytical index to an adapted bundle. Before developing the properties of this index in the next Section we want to interpose two remarks which will be needed in Chapter 4. First, the arguments in the proof of Proposition 3.6, while they have been given for the elliptic operator $D_{\mathbf{A}}$, apply more generally. Consider in particular the full covariant derivative

$$d_{\mathbf{A}} : \Omega^0_X(\mathfrak{g}_P) \to \Omega^1_X(\mathfrak{g}_P),$$

and its formal adjoint $d^*_{\mathbf{A}}$. We claim that the image of the extension $d_{\mathbf{A}} : L^2_1 \to L^2$ is *closed* in L^2. In this 4-dimensional case we can deduce the result from Proposition 3.6, since the image of d_A is the image of $L^2_1(\Omega^0)$ under $D_{\mathbf{A}}$, the summand $L^2_1(\Omega^0)$ is obviously L^2_1 closed in $L^2_1(\Omega^0 \oplus \Omega^+)$ and Fredholm operators preserve closedness. On the other hand one can give a direct proof that the image of $d_{\mathbf{A}}$ is closed, permuting the ingredients used in the proof of Proposition 3.6 but without passing through $D_{\mathbf{A}}$. This is obviously desirable – for example when considering manifolds of other dimensions. First, it is clear that the kernel of $d_{\mathbf{A}}$ on sections is zero, since the connection is assumed to be irreducible. Then the closedness of $\operatorname{Im} d_{\mathbf{A}}$ will follow easily (by considering Cauchy sequences) from a *Poincaré inequality*

$$\|f\|_{L^2} \leq C \|d_{\mathbf{A}} f\|_{L^2}.$$

The reader may find it a useful exercise to establish such an inequality, using separation of variables over the tubular ends and standard local elliptic theory in the interior.

As a corollary of the closedness of $\operatorname{Im} d_{\mathbf{A}}$ we obtain

Corollary 3.7 *If* $\mathbf{A}$ *is an adapted connection over* X *with acyclic limits there is an* L^2*-orthogonal decomposition*

$$L^2(\Omega^1_X(\mathfrak{g}_P)) = (\operatorname{Im} d_{\mathbf{A}} : L^2_1 \to L^2) \oplus \ker d^*_{\mathbf{A}}.$$

(Yet another approach to this would be to develop the Fredholm theory of the Laplacian $\Delta_{\mathbf{A}}$.)

The other remark we wish to place here is that the Fredholm theory applies to connections which are not precisely flat over the ends. Suppose as above that $\mathbf{P}$ is an adapted bundle and $\mathbf{A}$ an adapted connection. Let $\mathbf{A}' = \mathbf{A} + \mathbf{a}$ be another connection on $\mathbf{P}$ and suppose that $\mathbf{a}$ lies in C_0, the space of continuous sections with $|\mathbf{a}|$ tending to zero at infinity in X. Then $D_{\mathbf{A}'}$ maps L^2_1 to L^2, where the spaces are defined using $\mathbf{A}$ as before, and we claim that it is a Fredholm operator. Suppose first that $|\mathbf{a}|$ is small over the end $Y \times (0, \infty)$ in X, compared with the constant δ which depends only on the flat limit over Y. We consider an operator $D' = D + \mathbf{a}'$ over the full tube $Y \times \mathbf{R}$, with $\mathbf{a}'$ defined by extending the restriction of $\mathbf{a}$ by zero. If $|\mathbf{a}'|$ is small this is a small perturbation of the operator D, in the operator norm from L^2_1 to L^2, and so D' is invertible. Similarly, by the remarks after the proof of Lemma 3.2 above, we have for any solution f of the equation $D'f = 0$ over the half-tube, i.e. $Df = \rho = -\mathbf{a}' f$,

$$\begin{aligned} \|f\|_{L^2} &\leq \text{const.}\,(\|f|_B\|_{L^2} + c_\delta^{1/2}\|\rho\|_{L^2}) \\ &\leq \text{const.}\,(\|f|_B\| + (c_\delta^{1/2}\sup|\mathbf{a}'|)\|f\|). \end{aligned}$$

So if $\sup|\mathbf{a}'|$ is sufficiently small we get a bound on the L^2 norm of f from that of the restriction of f to B. Then, starting with these two ingredients, the proof of Proposition 3.6 shows that $D_{\mathbf{A}'}$ is Fredholm.

Finally if $\mathbf{a}$ is in C_0 we can always reduce to the situation considered above by redefining the parameter on the end of X – i.e. by considering the manifold to be a union of a compact piece and ends of the form $Y \times (T, \infty)$ for large T, in the original parametrisation.

In Section 3.3 below we will see that a Sobolev embedding $L_1^2 \subset L^4$ holds in this situation. It follows then from the argument above that $D_{\mathbf{A}'}$ is Fredholm if $\mathbf{a} = \mathbf{a}_1 + \mathbf{a}_2$ where $\mathbf{a}_1 \in C_0$ and $\mathbf{a}_2 \in L^4$.

3.3 The addition property

The index invariant defined in the previous Section has a simple formal property which is basic to Floer's theory. Let $\mathbf{A}$ be an adapted connection (with acyclic limits) on a bundle $\mathbf{P}$ over a 4-manifold X with tubular ends, as considered above, and suppose that X contains two boundary components $Y, \overline{Y}$, where $\overline{Y}$ is isometric to Y with the reversed orientation. More precisely, X has an end $Y \times (0, \infty)$ and another end $\overline{Y} \times (0, \infty)$. Suppose that the limiting flat connections appearing over Y and $\overline{Y}$ are the same. We fix an isometry between Y and $\overline{Y}$ and consider the family of Riemannian 4-manifolds $X^{\sharp(T)}$, depending on a real parameter $T > 0$, obtained by identifying the two ends of X. For fixed T we first delete the infinite portions $(Y \times [2T, \infty), \overline{Y} \times [2T, \infty)$ from the two ends, and then identify $(y, t) \in Y \times (0, T) \subset X$ with $(y, 2T - t) \in \overline{Y} \times (0, T) \subset X$. This gives a Riemannian manifold $X^{\sharp(T)}$, with two fewer ends than X. Clearly these are all diffeomorphic for different values of T. We will denote the manifolds by $X^\sharp$ when the T dependence is not important.

Suppose that the flat limits of $\mathbf{A}$ over Y and $\overline{Y}$ are the same, and fix an identification between these flat bundles. There is then an obvious way of constructing an adapted bundle $P^\sharp$ over $X^{\sharp(T)}$, using the flat structures to identify the bundles over the ends, and the adapted connection A gives a natural connection $A^\sharp$ on $P^\sharp$. We know that the index of the operator $D_{A^\sharp}$, which we denote by $\mathrm{ind}(P^\sharp)$, over $X^\sharp$ is independent of T, and the result we wish to prove in this section is the simple formula:

Proposition 3.8 *If X and $X^\sharp$ are 4-manifolds as above, and $P^\sharp$ is obtained from an adapted bundle P over X with acyclic limits then*

$$\mathrm{ind}(P^\sharp) = \mathrm{ind}(P).$$

The most important case of this for us will be when X is disconnected, say a disjoint union of two components $X = X_1 \cup X_2$, and the two ends which are identified are contained in different components of X. Then $X^{\sharp(T)}$ is connected. The procedure is a generalisation of the connected sum operation on manifolds (which occurs when Y is a 3-sphere). A connection A over X is a pair of connections A_1, A_2 on bundles P_1, P_2 over X_1, X_2. The index of the operator D_A over X is trivially the sum of the indices for the two components, so we obtain the additivity of the index with respect to generalised connected sums:

Proposition 3.9 *In this situation*

$$\mathrm{ind}(P^\sharp) = \mathrm{ind}(P_1) + \mathrm{ind}(P_2).$$

Although Proposition 3.8 is more general than Proposition 3.9, we will give the proof of this second result, since the notation is slightly simpler in this case, while the proof is identical in all essentials. The proof goes by analysing the operator $D_{A^\sharp}$ over $X^{\sharp(T)}$ for large values of T, and comparing its kernel and cokernel with those for the operators over the constituent manifolds. For $S \leq 2T$ let $X_i(S)$ denote the open set in X_i obtained by deleting the portion $Y \times [S, \infty)$ from the end. Then $X_i(S)$ can also be regarded as an open set in $X^{\sharp(T)}$, in an obvious way, and similarly functions etc. over X_i which are supported in $X_i(S)$ can be regarded as functions on $X^{\sharp(T)}$ (extending by zero). To simplify notation we write D_i for D_{A_i} and $D^\sharp$ for $D_{A^\sharp}$. The proof involves four steps; for the first three steps we suppose that the operators D_i over X_i have zero cokernel, so admit bounded right inverses

$$Q_i : L^2 \to L^2_1$$

with $\|Q_i(\rho)\|_{L^2} \leq C\|\rho\|_{L^2}$, say, and $D_i Q_i = 1$.

Step 1 In the first step we construct, for large T, an injection

$$\alpha : \ker D^\sharp \to \ker D_{A_1} \oplus \ker D_{A_2}.$$

In fact we construct a map α which is close to being an isometric embedding, with respect to the metrics on the kernels induced by the L^2 norms. To do this we fix functions ϕ_1, ϕ_2 on $X^{\sharp(T)}$ such that $\phi_1^2+\phi_2^2 = 1$, with ϕ_i supported in $X_i(3T/2)$ and such that $\|\nabla\phi_i\|_{L^\infty} = \epsilon(T)$, where $\epsilon(T) \to 0$ as $T \to \infty$. It is easy to write down such functions, indeed we can obviously take $\epsilon(T) = \text{const.}\, T^{-1}$. We then put, for $f \in \ker D^\sharp$, $\alpha(f) = (f_1, f_2)$ where

$$f_i = \phi_i f - Q_i D_i(\phi_i f).$$

Here we are regarding $\phi_i f$ as being defined over X_i in the obvious way, using the fact that ϕ_i is supported in $X_i(2T)$. The section f_i lies in the kernel of D_i, since Q_i is a right inverse. It is also a small perturbation of $\phi_i f$ in that we have

$$\begin{aligned} \|f_i - \phi_i f\| &= \|Q_i D_i(\phi_i f)\| \\ &\leq C\|D_i(\phi_i f)\| \\ &\leq C\||\nabla\phi_i| f\| \\ &\leq C\epsilon(T)\|f\|. \end{aligned}$$

Here we have used the fact that $D^\sharp f = 0$, and that D_i can be identified with $D^\sharp$ over the support of $\nabla\phi_i$. To complete the first step we now observe that, for any f on $X^{\sharp(T)}$,

$$\|\phi_1 f\|^2 + \|\phi_2 f\|^2 = \|f\|^2,$$

since $\phi_1^2 + \phi_2^2 = 1$. Otherwise said, the map $f \mapsto (\phi_1 f, \phi_2 f)$ defines an isometric embedding of $L^2{}_{X^{\sharp(T)}}$ in $L^2{}_{X_1} \oplus L^2{}_{X_2}$. This means that α is approximately an isometry for large T: precisely,

$$\begin{aligned} |\|\alpha(f)\| - \|f\|| &= |\|\alpha(f)\| - \|(\phi_1 f, \phi_2 f)\|| \\ &\leq \|(f_1 - \phi_1 f, f_2 - \phi_2 f)\| \\ &\leq \sqrt{2}C\epsilon(T)\|f\|, \end{aligned}$$

so α is injective once $\epsilon(T) < (1/\sqrt{2}C)$.

Step 2 For the second step we show that, under the same assumption of the existence of the right inverses Q_i, the operator $D^\sharp$ is also surjective for large T. To do this it suffices to construct a map $P : L^2 \to L^2_1$ over $X^{\sharp(T)}$ such that

$$\|D^\sharp P(\rho) - \rho\| \leq k\|\rho\|,$$

where $k < 1$. For then the operator $PD^\sharp - 1$ is invertible and $Q = P(PD^\sharp - 1)^{-1}$ is a right inverse for $D^\sharp$. We construct P by splicing together the operators Q_i over the individual manifolds. Write $\beta_i = \phi_i^2$, where ϕ_i are the cut-off functions above. Thus $\beta_1 + \beta_2 = 1$. Note that, since $0 \leq \phi_i \leq 1$, the gradient of β_i is bounded by $2\epsilon(T)$. We now follow the procedure used in the proof of (ii) in Proposition 3.6 – we define

$$P(\rho) = \beta_1 Q_1(\rho_1) + \beta_2 Q_2(\rho_2)$$

over $X^{\sharp(T)}$, where ρ_i is the restriction of ρ to $X_i(2T) \subset X^{\sharp(T)}$, extended by zero over the remainder of X_i. Similarly, $\beta_i Q_i(\rho_i)$ is regarded as a section over $X^{\sharp(T)}$, extending by zero outside the support of β_i. Then

$$\begin{aligned} D^\sharp P(\rho) &= (\beta_1 D^\sharp Q_1(\rho_1) + \beta_2 D^\sharp Q_2(\rho_2)) \\ &\quad +((\nabla\beta_1) * Q_1(\rho_1) + (\nabla\beta_2) * Q_2(\rho_2)). \end{aligned}$$

Here, as before, we use $*$ to denote a certain algebraic operation (essentially the symbol of the operator). Now $\beta_i D^\sharp Q_i(\rho_i) = \beta_i\rho_i = \beta_i\rho$, since we can identify $D^\sharp$ with D_i and ρ_i with ρ over the support of β_i. So the first two terms in the expression above yield ρ and the remainder has norm bounded by

$$\sum_{i=1}^{2} \|(\nabla\beta_i) * Q_i(\rho_i)\| \leq 4C\epsilon(T)\|\rho\|.$$

So $\|D^\sharp P\rho - \rho\| \leq 4C\epsilon(T)\|\rho\|$, and we achieve the desired 'approximate inverse' by taking T so large that $\epsilon(T) \leq 1/4C$. This completes the second step in the proof.

Step 3 In the third step we construct, under the same assumption of the surjectivity of D_i, a linear injection $\alpha' : \ker D_1 \oplus \ker D_2 \to \ker D^\sharp$ for large enough T. For this we first return to the construction of the operator Q above, and note that it admits an L^2 bound:

$$\|Q(\rho)\| \leq 3C\|\rho\|,$$

say, for all large enough values of T. For the operator norm of P is clearly at most $2C$, and we can make the norm of $(1 - D^\sharp Q)^{-1}$ as close to 1 as we please. With this observation the map α' can be constructed in a similar fashion to the map α in the first step. For elements f_i of the kernels of the D_i over X_i we set $\alpha'(f_1, f_2) = g - QD^\sharp g$, where

$$g = \beta_1 f_1 + \beta_2 f_2.$$

Here we have identified appropriate sections over X_i and X, in the way which will now be familiar to the reader. Just as in the first step, we see that the L^2 norm of the 'correction term' $QD^\sharp g$ is bounded by an arbitrarily small multiple of $\|(f_1, f_2)\|$. It remains only to show that the L^2 norm of g is close to that of (f_1, f_2) for large T. Let f^σ, $\sigma = 1, \ldots, d$, be an orthonormal basis for the fixed space $\ker D_1$. Then $F = \sum |f^\sigma|^2$ is an integrable function of X_1. For any $\eta > 0$ we can thus choose T_0 so that, in the end $Y \times [0, \infty) \subset X_1$,

$$\int_{(T_0/2,\infty)} F \leq \eta.$$

For $T \geq T_0$ then,

$$\int_{(T/2,\infty)} |f_1|^2 \leq \eta \|f_1\|^2,$$

for any $f_1 \in \ker D_1$. We can suppose the same inequality over X_2 holds and then, since $\beta_i = 1$ on the segment $Y \times (0, T/2)$ of the tube in X_i, we clearly have

$$\|\beta_1 f_1 + \beta_2 f_2\|^2 \geq (1 - \eta) \|(f_1, f_2)\|^2,$$

so α' is almost an isometry.

These first three steps complete the proof of the 'gluing formula' (Proposition 3.9) in the case when the D_i are surjective. For in this case we have, by step 2, $\mathrm{ind}(X_i, P_i, A_i) = \dim \ker D_i$, $\mathrm{ind}(P^\sharp) = \dim \ker D^\sharp$ for large T. By step 1, $\dim \ker D^\sharp \leq \dim \ker D_1 + \dim \ker D_2$, and step 3 gives the reverse inequality, so $\dim \ker D^\sharp = \dim \ker D_1 + \dim \ker D_2$ as required. (It follows that α and α' are both isomorphisms, for large T. It is easy to see also that they are approximately inverse maps.)

Step 4 In the final step we remove the assumption that the operators D_i are surjective. We do this by modifying the operators. We can choose maps

$$U_i : \mathbf{R}^{n_i} \to (\Omega^0 \oplus \Omega^+_{X_i})(\mathfrak{g}_{P_i}),$$

with images supported in the interior of the X_i, and such that

$$\tilde{D}_i \equiv D_i \oplus U_i : \Omega^1 \oplus \mathbf{R}^{n_i} \to (\Omega^0 \oplus \Omega^+)$$

is surjective. The index of $\tilde{D}_i$ is $\mathrm{ind}(P_i)$ (consider the homotopy $D_i \oplus sU_i$, $0 \leq s \leq 1$, through Fredholm operators). We can form an obvious operator $\tilde{D^\sharp} = D^\sharp \oplus U_1 \oplus U_2$ over $X^{\sharp(T)}$, and the proof above goes over without any change to show that $\mathrm{ind}\ \tilde{D^\sharp} = \mathrm{ind}\, \tilde{D}_1 + \mathrm{ind}\, \tilde{D}_2$ so

$$\begin{aligned} \mathrm{ind}\, D^\sharp &= \mathrm{ind}\, \tilde{D^\sharp} - (n_1 + n_2) \\ &= (\mathrm{ind}\, \tilde{D}_1 - n_1) + (\mathrm{ind}\, \tilde{D}_2 - n_2) \\ &= \mathrm{ind}\, D_1 + \mathrm{ind}\, D_2, \end{aligned}$$

as required.

3.3.1 Weighted spaces

We will now extend the theory to include adapted bundles over 4-manifolds whose limits are not acyclic. Thus, going back to three dimensions, we have to consider a flat connection A over a 3-manifold Y for which 0 appears in the spectrum $\{\lambda\}$ of the operator L_A. We now let δ be the minimum absolute value of a *non-zero* eigenvalue of L.

Over the tube $Y \times \mathbf{R}$ it is not now true that $D = \frac{d}{dt} + L$ defines an invertible operator, or even a Fredholm operator, from L^2_1 to L^2. The trouble comes from the component

$$\frac{d}{dt} : L^2_1 \to L^2$$

in the spectral decomposition of D, corresponding to the zero eigenspace of L_A. It is easy to see that $\frac{d}{dt}$ is not a Fredholm operator from $L^2_1(\mathbf{R})$ to $L^2(\mathbf{R})$. For we can take a sequence of functions g_n on $\mathbf{R}$, with $g_n(t) = 1$ for $t \in (-n, n)$, supported in $(-(n+1), (n+1))$, with $\|\frac{d}{dt} g_n\| \leq$ const. and $\|g_n\| \to \infty$. If $\frac{d}{dt}$ were Fredholm this would imply that it had a non-trivial kernel in L^2_1, which is clearly not the case (the constants are not in L^2).

To recover a Fredholm theory we use weighted function spaces. For any α we define the weighted L^2 norm on the sections of $(\Lambda^0_Y \oplus \Lambda^1_Y)(\mathfrak{g}_P)$ over $Y \times \mathbf{R}$ by

$$\|f\|^2_{L^{2,\alpha}} = \int_{Y \times \mathbf{R}} e^{2\alpha t} |f|^2.$$

That is, $\|f\|_{L^{2,\alpha}} = \|e^{\alpha t} f\|_{L^2}$. Similarly, we define a weighted L^2_1 norm by

$$\|f\|_{L^{2,\alpha}_1} = \|e^{\alpha t} f\|_{L^2_1}.$$

(Note that this is equivalent to the norm $\|e^{\alpha t} \nabla f\| + \|e^{\alpha t} f\|$.)

We then define Hilbert spaces $L^{2,\alpha}, L^{2,\alpha}_1$ to be the completions of the smooth, compactly supported, sections in these norms. (The same argument as before shows that it is equivalent to define $L_1^{2,\alpha}$ as the sections $f \in L^{2,\alpha}$ with $Df \in L^{2,\alpha}$.) Roughly speaking, for $\alpha > 0$ these spaces admit functions which *decay* exponentially as $t \to \infty$ (and grow exponentially as $t \to -\infty$), and for $\alpha < 0$ we get exponential decay at $+\infty$ and growth at $-\infty$.

Now the deformation operator over $Y \times \mathbf{R}$ extends to a bounded operator

$$D = D_{\mathbf{A}} = \frac{d}{dt} + L_A : L_1^{2,\alpha} \to L^{2,\alpha}.$$

We claim that this operator is invertible for any weight α which does *not* lie in the spectrum of L_A. In particular this is true for non-zero α with $|\alpha| < \delta$. This follows easily from the previous result for the acyclic case. We only have to observe that multiplication by $e^{\alpha t}$ gives isometries from $L_1^{2,\alpha}$ to L_1^2 and $L^{2,\alpha}$ to L^2. Composing with these isometries we see that $D : L_1^{2,\alpha} \to L^{2,\alpha}$ is equivalent, as a map of Hilbert spaces, to

$$e^{\alpha t} D e^{-\alpha t} : L_1^2 \to L^2.$$

Now

$$\begin{aligned} e^{\alpha t} D e^{-\alpha t} &= e^{\alpha t} \left(\frac{d}{dt} + L \right) e^{-\alpha t} \\ &= \frac{d}{dt} + (L - \alpha). \end{aligned}$$

That is, the use of the weighted spaces is effectively the same as replacing L by $L - \alpha$. Now the previous results apply as well to $L - \alpha$ as to L, so long as it is an invertible operator over Y, i.e. so long as α is not in the spectrum of L, and we deduce that D is then invertible on the weighted spaces, just as before.

We can now go on to set up the index theory over a general 4-manifold X with tubular ends. We have to choose a weight α_i for each end $Y_i \times (0, \infty)$ of X. Fix a positive function W on X which is equal to $e^{\alpha_i t}$ on the ith end and define norms:

$$\|f\|_{L^{2,\underline{\alpha}}} = \|Wf\|_{L^2}, \quad \|f\|_{L_1^{2,\underline{\alpha}}} = \|Wf\|_{L_1^2}$$

with completions $L^{2,\underline{\alpha}}, L_1^{2,\underline{\alpha}}$. Different choices of W, with the same weight vector $\underline{\alpha} = (\alpha_1, \alpha_2, \ldots)$, give equivalent norms.

The argument used before goes through without change to show that $D : L_1^{2,\underline{\alpha}} \to L^{2,\underline{\alpha}}$ is Fredholm so long as α_i does not lie in the spectrum of the operator L over Y_i. We thus have an index:

$$\operatorname{ind}(P, \underline{\alpha}) = \operatorname{ind} D_{\mathbf{A}} : L_1^{2,\underline{\alpha}} \to L^{2,\underline{\alpha}},$$

for any compatible connection $\mathbf{A}$. It follows from the remarks in Section 2.1 that this index does not change if $\underline{\alpha}$ is varied in such a way that α_i avoids the spectrum of L over Y_i. Conversely, as we shall see below, the index *will* change if α_i is moved across an eigenvalue.

We begin by considering the gluing problem. Suppose we are in the situation of Proposition 3.8, with a 4-manifold X which contains $Y = Y_1$ and $\overline{Y} = Y_2$ among its ends, and we have an adapted connection $\mathbf{A}$ on a bundle $\mathbf{P}$ which agrees on these two cross-sections. Let $\underline{\alpha} = (\alpha_1, \alpha_2, \dots)$ be a vector of weights, and consider $D = D_{\mathbf{A}} : L_1^{2,\underline{\alpha}} \to L^{2,\underline{\alpha}}$. This is conjugate, by the function W, to an operator of the form $D + \nu : L_1^2 \to L^2$, where ν is an algebraic operator, represented as multiplication by α_i in our description over the ith end. The operator D is represented as $\frac{d}{dt_1} + L_Y$ over the first end and as $\frac{d}{dt_2} + L_{\overline{Y}}$ over the second, where now we write t_1, t_2 for the 'time' co-ordinates over the different ends. When we identify the ends to form $X^\sharp$ we reverse the time co-ordinates, so $\frac{d}{dt_1}$ corresponds to $-\frac{d}{dt_2}$, and this marries up with the natural identification $L_Y = -L_{\overline{Y}}$. (The apparent overall difference in sign between the operators is absorbed by the orientation conventions we have used in our representation of the operator over a tube.) Thus, in our gluing operation, the operator

$$\frac{d}{dt_1} + L_Y + \alpha_1$$

over the first end can naturally be identified with

$$\frac{d}{dt_2} + L_{\overline{Y}} + \alpha_2$$

over the second if

$$\alpha_1 = -\alpha_2.$$

In this case the arguments we used before go through without any change to show that

$$\operatorname{ind}(P^\sharp, (\alpha_3, \dots, \alpha_N)) = \operatorname{ind}(\mathbf{P}, (\alpha_1, -\alpha_1, \alpha_3, \dots, \alpha_N)). \tag{3.2}$$

We now consider the dependence of the index on the weights. To simplify notation we consider a situation where there is only one end and we compare weights α^+, α^-, where α^+ is slightly larger than zero (i.e. below the positive spectrum of L_Y) and α^- is slightly smaller than zero. So we have two indices

$$\mathrm{ind}^+(P) = \mathrm{ind}(P, \alpha^+), \quad \mathrm{ind}^-(P) = \mathrm{ind}(P, \alpha^-).$$

Proposition 3.10 *In this situation the index difference is*

$$\mathrm{ind}^+(P) - \mathrm{ind}^-(P) = -\dim\ker L_Y.$$

(Of course, in the notation of Chapter 2, the dimension of the kernel of L_Y is the sum of the dimensions of the cohomology groups H^0_ρ, H^1_ρ.)

To establish this formula we first use the gluing Equation 3.2 to reduce to the tube $Y \times \mathbf{R}$. We consider the deformation operator D_0 defined by the flat connection over $Y \times \mathbf{R}$, acting on spaces constructed from a weight function e^σ, with

$$\begin{aligned} \sigma(t) &= \alpha t \quad \text{for } t \gg 0, \\ &= -\alpha t \quad \text{for } t \ll 0, \end{aligned}$$

where α is a small positive number.

We apply Equation 3.2 to the disjoint union $X \cup (Y \times \mathbf{R})$. Gluing the tube to X, with the given weights, changes one index problem to the other one. So for a bundle P over X the gluing result gives

$$\mathrm{ind}^+(P) - \mathrm{ind}^-(P) = \mathrm{index}\ D_0.$$

To compute the index of D_0 on the spaces weighted by e^σ we use the same argument as before to see that this is the same as the index of

$$D_0' \equiv e^{\sigma(t)} D_0 e^{-\sigma(t)} = \frac{d}{dt} + (L - \sigma'(t)) : L^2_1 \to L^2.$$

Now the operation of multiplication by $\sigma'(t)$ commutes with L, so we can decompose D_0' according to the spectrum of L

$$D_0'\left(\sum_\lambda f_\lambda \phi_\lambda\right) = \sum_\lambda \left(\left(\frac{d}{dt} + \lambda - \sigma'(t)\right) f_\lambda\right) \phi_\lambda.$$

What we require, then, is the following simple result on ordinary differential equations.

Lemma 3.11 *Let τ be a smooth, real-valued function on $\mathbf{R}$, with $\tau(t) = \tau^+$ for $t \gg 0$ and $\tau(t) = \tau^-$ for $t \ll 0$, where τ^+, τ^- are non-zero. Define K^+, K^- to be the space of L^2 solutions to the equations*

$$\left(\frac{d}{dt} + \tau(t)\right) f = 0, \quad \left(-\frac{d}{dt} + \tau(t)\right) f = 0$$

respectively. Then

- *if* τ^+, *and* τ^- *have the same sign then* K^+ *and* K^- *are both zero,*
- *if* $\tau^+ > 0$ *and* $\tau^- < 0$ *then* $K^+ = \mathbf{R}$ *and* $K^- = 0$,
- *if* $\tau^+ < 0$ *and* $\tau^- > 0$ *then* $K^+ = 0$ *and* $K^- = \mathbf{R}$.

The reader will have no difficulty in supplying a proof of this Lemma. We apply the results to the co-efficients $f_{\lambda(t)}$, with $\tau = \lambda - \sigma'$, to deduce that, for an element f of the kernel of D'_0, the co-efficients f_λ vanish for all λ; so the kernel is trivial. Similarly we see the adjoint of D'_0 has a kernel isomorphic to $\ker L_Y$, so the index of D'_0 is $-\dim\ker L$ as required.

To round off this discussion of the index theory we consider two special cases. One case is when the ends of the 4-manifold X have spheres for cross-sections, so there is a smooth compactification $\hat{X}$. If $\mathbf{P}$ is an adapted bundle with trivial limit over the ends it extends naturally to a bundle $\hat{\mathbf{P}}$ over $\hat{X}$. The other special case is when the bundle $\mathbf{P}$ over the 4-manifold is trivial; so we may reduce to the operator $D = d^* \oplus d^+$ on ordinary forms. We begin with the model case of the 'flask' manifold W, obtained by adding a punctured 4-sphere to a tube $S^3 \times \mathbf{R}^+$. Let ind^+_W, ind^-_W denote the indices for the operator D coupled to the trivial bundle (of rank 3), with small positive and negative weights respectively. We have

Lemma 3.12 *The indices for* W *are* $\mathrm{ind}^+_W = -3$, $\mathrm{ind}^-_W = 0$

Observe that W admits an orientation reversing isometry and the double $W \sharp W$ is diffeomorphic to S^4. We know that the index of the operator D over S^4 is -3 so the gluing rule gives $\mathrm{ind}^+_W + \mathrm{ind}^-_W = -3$. On the other hand by Proposition 8,

$$\mathrm{ind}^+_W = \mathrm{ind}^-_W - 3$$

and we are done.

We have then

Corollary 3.13 *If the ends of* X *have the form* $S^3 \times \mathbf{R}$ *and the limits of* $\mathbf{P}$ *are trivial then*

$$\mathrm{ind}^+(\mathbf{P}) = \mathrm{ind}(\hat{\mathbf{P}}) = 8c_2(\hat{\mathbf{P}}) - 3(1 - b_1(X) + b_+(X)).$$

To prove this we cap off each end with a copy of the flask W, and use the gluing relation.

We now turn to the second special situation, when the bundle is trivial but we allow any tubular ends. It will come as no surprise that the index in this situation can be related to cohomology via a version of the Hodge theory. We can do more and look at the actual kernels of the operators D and D^* on the weighted spaces. We let $\ker_+$ denote the kernel on the spaces with a small positive weight (i.e. the decaying solutions) and $\ker_-$ denote that with a negative weight (i.e. allowing bounded solutions). Here one should remember that the kernel of the formal adjoint D^* considered on the spaces with negative weights is isomorphic to the cokernel of D considered on the spaces with positive weights, and vice versa. We consider first a simple situation when X has one end which has a homology 3-sphere Y as cross-section. Then X behaves like a closed manifold as far as cohomology goes (more precisely, the compactification $\hat{X}$ is a homology manifold), in particular the cup product defines a non-degenerate quadratic form on the real cohomology group $H^2(X)$.

Proposition 3.14 *When the cross-section of the end is a homology 3-sphere we have*

$$\ker_+ D = \ker_- D \cong H^1(X),$$

and

$$\ker_+ D^* = H^+(X),\ \ker_- D^* = H^0(X) \oplus H^+(X),$$

where $H^+(X) \subset H^2(X)$ is a maximal positive subspace for the cup product (intersection) form.

Notice here that the adjoint of D on the positively weighted spaces is the operator D^* on the *negatively* weighted spaces, so the index of D on the positively weighted spaces is $b_1 - (1 + b_+)$, in agreement with Corollary 3.13.

The proof is a straightforward exercise, most of which we leave to the reader. To define a map from $\ker_+ D$ to $H^1(X)$, for example, one needs to verify that if α is an L^2_1 1-form over X with $d^+\alpha = 0$ then $d\alpha = 0$. This follows by considering integrals

$$\int_X d(\beta_T \alpha \wedge d\alpha)$$

where β_T are cut-off functions as in Subsection 2.1.1.

The formulae for a general manifold X, with boundary Y, are slightly more complicated. We now have absolute and relative cohomology groups $H^*(X)$, $H^*(X,Y)$, related by Poincaré duality and to $H^*(Y)$ by the usual long exact sequence. There is a quadratic form on the image, I say, of the map $H^2(X,Y) \to H^2(X)$ and, even though this may be degenerate, the notion of a maximal positive subspace makes sense.

Proposition 3.15

- $\ker_+ D \cong \operatorname{Im} H^1(X,Y) \to H^1(X)$,
- $\ker_- D \cong H^1(X)$,
- $\ker_+ D^* \cong H^+(X)$,
- $\ker_- D^* \cong H^0(X) \oplus H^+(X)$,

where $H^+(X) \subset I$ is a maximal positive subspace for the cup product form.

Again, we leave the proof as an exercise for the reader.

3.3.2 *Floer's grading function; relation with the Atiyah, Patodi, Singer theory*

The discussion of the previous Subsection has shown that to any adapted $SU(2)$ bundle $\mathbf{P}$ over a 4-manifold X with tubular ends, which is acyclic over each end, we can associate an integer invariant $\operatorname{ind}(\mathbf{P})$. If the limits are not acyclic we can associate a pair of invariants $\operatorname{ind}^+(\mathbf{P})$, $\operatorname{ind}^-(\mathbf{P})$ whose difference is the dimension of the kernel of the L operators over the ends. Suppose that $\mathbf{P}'$ is another adapted bundle over X with the same flat limits as $\mathbf{P}$. We have seen that $\mathbf{P}, \mathbf{P}'$ need not be isomorphic as adapted bundles and the distinction between them is precisely measured by the difference in the relative Chern classes

$$c_2(\mathbf{P}') - c_2(\mathbf{P}) \in \mathbf{Z}.$$

Proposition 3.16 *If the limits are acyclic then*

$$\operatorname{ind}(\mathbf{P}') - \operatorname{ind}(\mathbf{P}) = 8(c_2(\mathbf{P}') - c_2(\mathbf{P})),$$

and in general $\operatorname{ind}^{\pm}(\mathbf{P}') - \operatorname{ind}^{\pm}(\mathbf{P}) = 8(c_2(\mathbf{P}') - c_2(\mathbf{P}))$.

This is a rather routine consequence of the index formula for closed manifolds, using an excision argument. In our present context it is natural to use the gluing Equation 3.2. We puncture X and modify the metric to make a new manifold X_0 with one extra end, having a 3-sphere as cross-section. Then we can recover X by gluing X_0 to another manifold, the flask W of Lemma 3.12, in the obvious way. It is clear that we can obtain $\mathbf{P}$ and $\mathbf{P}'$ by gluing a bundle $\mathbf{P}_0$ over X_0 to suitable adapted bundles $\mathbf{P}_1, \mathbf{P}'_1$ over W. Of course the flat limit of the bundles over the end of W is necessarily trivial and hence not acyclic. However, we can apply our gluing relation Equation 3.2. The Chern class is plainly additive,

$$c_2(\mathbf{P}) = c_2(\mathbf{P}_0) + c_2(\mathbf{P}_1),$$

so, using the gluing relation for the index, it suffices to show that, over X_1,

$$\mathrm{ind}^+(\mathbf{P}_1) - \mathrm{ind}^+(\mathbf{P}'_1) = 8c_2(\mathbf{P}_1) - c_2(\mathbf{P}'_1)).$$

But, by Corollary 3.13, the index with positive weights ind^+, in the case when the ends are spheres, agrees with the usual index of the operator we obtain by compactifying the manifold. Then the result follows immediately from the index Formula .2.5 for the base manifold S^4.

Consider in particular the case when the base manifold is a tube $Y \times \mathbf{R}$. For any pair of acyclic flat connections ρ, ρ' over Y we can define a *relative index*

$$\delta_Y(\rho, \rho') \in \mathbf{Z}/8$$

as follows. We choose a bundle $\mathbf{P}$ over the tube with limits ρ at $+\infty$ and ρ' at $-\infty$ and we set

$$\delta_Y(\rho, \rho') = \mathrm{ind}(\mathbf{P}) \mod 8. \tag{3.3}$$

By Proposition 3.16 this definition does not depend on the bundle $\mathbf{P}$ used. This relative index is a crucial ingredient in Floer's theory – we shall see that it will induce the grading in the Floer homology groups. In the case of $SU(2)$ bundles, which we are considering here, it can be refined slightly to a $(\mathbf{Z}/8)$-grading function δ_Y on the acyclic flat connections (as we mentioned at the end of Chapter 2). To define this we use the trivial flat connection θ as a reference point. For an acyclic connection ρ we choose a bundle $\mathbf{P}$ over the tube with flat limits ρ at $+\infty$ and θ at $-\infty$. Then we put

$$\delta_Y(\rho) = \mathrm{ind}^-(\mathbf{P}) \mod 8. \tag{3.4}$$

Our addition relation for the index now gives

$$\delta_Y(\rho, \rho') = \delta_Y(\rho) - \delta_Y(\rho').$$

More generally we have

Proposition 3.17 *Let X be a 4-manifold with one tubular end $Y \times \mathbf{R}^+$, where Y is a homology 3-sphere, and let $\mathbf{P}$ be an adapted $SU(2)$ bundle over X with acyclic limit ρ. Then*

$$\operatorname{ind}(\mathbf{P}) = \delta_Y(\rho) + 3(1 - b_1(X) + b_+(X)) \mod 8.$$

This is a straightforward consequence of the gluing relation and the Hodge theory (Proposition 3.14).

The ideas we have been discussing are closely related to the index theory for manifolds with boundary developed by Atiyah, Patodi and Singer in their series of papers beginning with [4]. In particular there are two concepts from this theory – the *eta-invariant* and the *spectral flow* of a family – which it is natural to mention here, although they will not be used later in this book. We begin with spectral flow. Observe first that throughout this Chapter there was no need to restrict to connections with flat limits. We could more generally consider any connection on a bundle over a 4-manifold with tubular ends which is equal to a pull-back $\pi^*(A)$ of a connection A from Y over an end $Y \times \mathbf{R}^+$. We call the connection A acyclic if the operator L_A over Y is invertible and we define an index $\operatorname{ind}(P)$ in the acyclic case and weighted indices $\operatorname{ind}^{\pm}$ in the general case. Thus we can extend our grading function δ_Y to the open set $U \subset \mathcal{B}_P$ of all acyclic connections. The picture that emerges is that U is partitioned by δ into eight pieces $U_0, \dots, U_7$, separated by codimension-1 'walls' of non-acyclic connections. A typical one-parameter family of connections A_s, $s \in [0,1]$, will meet the walls in isolated points – the points where $L_s \equiv L_{A_s}$ acquires a zero eigenvalue. We can think of the spectrum of this family as a 1-dimensional subset of $[0,1] \times \mathbf{R}$. The spectral flow of the family is defined to be the algebraic intersection number of the spectrum with the zero axis – i.e. we count an eigenvalue that increases through zero with a $+1$ and one that decreases with a -1. (To interpret this more precisely one may need to perturb to a sufficiently generic situation.) This gives another way of viewing the relative index in that we have

Proposition 3.18 *Let A_0 and A_1 be two acyclic connections and A_s be a path between them. Then $\delta_Y(A_0, A_1)$ is equal,* mod 8, *to the spectral flow of the path of operators L_s over Y.*

This can easily be deduced form the theory above, and we sketch the argument. First, by continuity, the result is true if there are no zero eigenvalues anywhere in the path. We can then use the additive properties to reduce to a case when there is just one eigenvalue crossing and this is transverse, increasing say. We extend to an infinite path in the obvious way and consider the operator $D = \frac{d}{dt} + L_t$ over the tube. We then manufacture a family of the form $\tilde{L}_t = e^\sigma L_1 e^\sigma$ with the opposite flow, taking a weight function σ equal to 1 for $t \ll 0$ and to the exponential of ϵt for $t \gg 0$, for a suitable ϵ. This gives an operator $\tilde{D} = \frac{d}{dt} + \tilde{L}_t$. We can adapt our gluing formula to show that the index of the 'connected sum' of D and $\tilde{D}$ is the sum of the individual indices. Since we can calculate the latter index explicitly as in Lemma 3.11, it suffices to show that the index of the connected sum is 0, and this is done by perturbing by the addition of a positive scalar to shift the spectrum away from the zero axis.

We now turn to the Atiyah, Patodi and Singer eta-invariant. This arises when one attempts to generalise integral formulae for the index, in the closed case, to manifolds with tubular ends. Let $\mathbf{A}$ be a connection on a bundle $\mathbf{P}$ over a 4-manifold with a tubular end, which is equal to the constant acyclic connection A over the end. Then the Chern–Weil integrand $\mathrm{Tr}(F_{\mathbf{A}}^2)$ is compactly supported. Combining the theory of Chapter 2 and this Chapter, we know that the real number

$$\gamma(A) = \mathrm{ind}(P) - \frac{1}{8\pi^2}\int_X \mathrm{Tr}(F_{\mathbf{A}}^2) - 3(1 - b_1(X) + b_+(X))$$

gives a lift of the Chern–Simons invariant $\vartheta(A)$ to $\mathbf{R}$. The Atiyah, Patodi and Singer theory gives a *spectral formula* for this number

$$\gamma(A) = \eta_A(0) - \eta_\theta(0) + 3/2.$$

Here $\eta_A(0)$ is defined as follows. For $\Re s \gg 0$ the Dirichlet series formed from the spectrum$\{\lambda\}$ of L_A,

$$\eta_A(s) = \sum_\lambda \mathrm{sgn}(\lambda)|\lambda|^{-s},$$

converges to define an analytic function $\eta_A(s)$. This can be analytically continued to a neighbourhood of 0 and $\eta_A(0)$ is the value so obtained. The next term $\eta_\theta(0)$ in the formula is obtained in the same way using the trivial flat connection θ, except one must then remove the zero eigenvalue from the series. The final term 3/2 appears as a correction term due to the 3-dimensional zero eigenspace.

3.3.3 Refinement of weighted theory

The version of the weighted theory that we shall actually need in Chapter 4 is a little different from that above. Recall that we introduced the operator $D_A = -d_A^* + d_A^+$ in four dimensions as a tool to study the deformation complex. Thus the two parts d_A^* and d_A^+ have a quite distinct geometrical significance, although the discussion so far throughout this Chapter treats them on an equal footing. The refinement we discuss now brings in the separate nature of these two components.

Suppose X is a 4-manifold with a tubular end, and W is a weight function equal to $e^{\alpha t}$ over the end, where α is positive. We fix a connection on an adapted bundle P as before, with flat limit ρ. We have then an operator, acting on sections of $\mathfrak{g}_P$,

$$d_A : L_1^{2,\underline{\alpha}} \to L^{2,\underline{\alpha}}.$$

We consider the adjoint operator, with respect to the *weighted* norms. That is, the operator $d_A^{*,\underline{\alpha}}$ defined by the condition that

$$\int_X W^2(d_A f, a) = \int_X W^2(f, d_A^{*,\underline{\alpha}} a),$$

for all f and a of compact support. Clearly this operator is just

$$d_A^{*\underline{\alpha}}(a) = W^{-2} d_A^*(W^2 a).$$

In particular, over the end of X we can write

$$d_A^{*,\underline{\alpha}} = d_A^* + 2\alpha i_t(a)$$

(where i_t is the contraction with the vector $\frac{d}{dt}$). Now let us consider the operator

$$D_{A,\underline{\alpha}} = -d_A^{*,\underline{\alpha}} + d_A^+,$$

acting on the $\mathfrak{g}_P$-valued 1-forms in $L_1^{2,\underline{\alpha}}$ over X, i.e.

$$D_{A,\underline{\alpha}} : L_1^{2,\underline{\alpha}} \to L^{2,\underline{\alpha}}.$$

Proposition 3.19 *For small positive α the operator $D_{A,\underline{\alpha}}$ is Fredholm of index*

$$\operatorname{ind} D_{A,\underline{\alpha}} = \operatorname{ind}^+(P) + \dim H_\rho^0.$$

This can be seen as a variant of Proposition 3.9 which says that the jump in the index from ind^+ to ind^- is the sum of the dimensions of the cohomology groups H_ρ^0, H_ρ^1: the operator $D_{A,\underline{\alpha}}$ allows us to the see the two phenomena separately.

The proof of Proposition 3.19 follows the same lines as that of Proposition 3.9. Consider the operator $D_{A,\underline{\alpha}}$ over the tube, acting on weighted spaces with weight function $e^{\alpha t}$. As before, this is equivalent to an operator, D'' say, acting on unweighted spaces, where

$$D''(a) = e^{\alpha t} D_{A,\underline{\alpha}}(e^{-\alpha t} a) = D_{A,\underline{\alpha}} - \alpha.$$

Thus $D''(a) = D_A + 2\alpha i_t - \alpha$. Now recall that, over the tube, D_A is regarded as acting on the sections of the direct sum $\Lambda^0_Y \oplus \Lambda^1_Y$ (tensored with the flat bundle). Let R be the linear operator that acts as $+1$ on Λ^0_Y and -1 on $\Lambda^1(Y)$. Then we have

$$D'' = D_A + \alpha R = \frac{d}{dt} + L_Y + \alpha R.$$

The essential thing, therefore, is to consider the spectrum of $L_Y + \alpha R$ on sections of $\Lambda^0_Y \oplus \Lambda^1_Y$ over the 3-manifold Y. Recall that in Subsection 2.5.4 we decomposed L_Y as a sum $L_Y = Q \oplus S$ where Q acts on $\ker d^* \subset \Omega^1$ and S acts on $\Omega^0 \oplus \operatorname{Im} d \subset \Omega^1$ by the matrix

$$\begin{pmatrix} 0 & -d \\ -d^* & 0 \end{pmatrix}.$$

The eigenvalues of S have the form

$$\begin{pmatrix} \pm\sqrt{\mu} d\psi_\mu \\ \psi_\mu \end{pmatrix}$$

where ψ_μ is an eigenfunction of the Laplacian on 0-forms with eigenvalue μ. Now the operator $L_Y + \alpha R$ also decomposes into two pieces. One piece is $Q - \alpha$, so the spectrum of this piece is just the spectrum of Q shifted by $-\alpha$. The other piece can be written as

$$\begin{pmatrix} -\alpha & -d \\ -d^* & \alpha \end{pmatrix}.$$

It is straightforward to check then that the spectrum of this second piece is the set of numbers of the form $\pm\sqrt{\mu + \alpha^2}$, where μ runs over the spectrum of the Laplacian, but in the case of a zero eigenvalue, $\mu = 0$, we take only the *positive* square root. Thus, if α is small and positive, the effect of the replacing of S by $S + \alpha R$ is to shift the eigenvalues in the *positive direction* and in particular to replace the zero eigenvalue by α.

Now to prove Proposition 3.19 we consider an operator over the tube of the form

$$D_A - \alpha + \sigma(t) i_t,$$

where the function $\sigma(t)$ is zero for $t \ll 0$ and equal to 2 for $t \gg 0$. The same spectral decomposition argument shows that the index of this operator is the dimension of H^0_ρ (i.e. the multiplicity of the eigenvalue $\mu = 0$). Then the familiar gluing argument and the additivity property give Proposition 3.19.

3.4 L^p theory

When we develop the non-linear analysis of Yang–Mills fields on manifolds with tubular ends in Chapter 4, we will need other function spaces, and appropriate extensions of the results above. These are the spaces L^p_k of sections with 'k derivatives in L^p'. As usual, we concentrate here on the changes that need to be made in going from compact manifolds to manifolds with tubular ends. We begin by reviewing the main results which hold over compact sets. First, over any space of finite measure the L^p spaces are naturally ordered by inclusion: for $p > q$ the L^p norm is strictly stronger than the L^q norm. The *Sobolev inequalities* allow us to increase the strength of the exponent defining the norm in exchange for taking derivatives. If U is a compact Riemannian manifold of dimension d, possibly with boundary, for any $p, r > 1$ there is a constant $c_{U,p,r}$ such that for any section f of a bundle over U (with metric and unitary connection),

$$\begin{aligned} \|f\|_{L^q} &\equiv \left(\int_U |f|^q\right)^{1/q} \leq c_{U,p,r}\left(\left(\int_U |\nabla f|^p\right)^{1/p} + \left(\int_U |f|^r\right)^{1/r}\right) \\ &\equiv c_{U,p,r}(\|\nabla f\|_{L^p} + \|f\|_{L^r}), \end{aligned} \tag{3.5}$$

where the indices p and q are related by

$$1 - \frac{d}{q} = -\frac{d}{p}. \tag{3.6}$$

The second basic local result allows us to exchange the full covariant derivative ∇ for an elliptic operator D. If $U' \subset\subset U$ is a subdomain we have

$$\left(\int_{U'} |\nabla f|^p\right)^{1/p} \leq c'_{U,U',p,r}\left(\int_U |Df|^p\right)^{1/p} + \left(\int_U |f|^r\right)^{1/r}. \tag{3.7}$$

This inequality, for general p, is derived from the comparatively deep results of the Calderon–Zygmund theory [43]. Given these L^p results for compact sets it is, however, straightforward to extend them to manifolds with tubular ends. In effect when we pass to a non-compact manifold the interesting new features, beyond the local theory, involve decay at infinity. But in controlling this decay the L^q norm is *weaker* than the L^p norm, if $q > p$, so the L^p results (at least the results we shall consider) have rather little to do with the special 'non-compact' features of the theory.

We begin with the tube $Y \times \mathbf{R}$ and suppose, as in Section 3.1, that D is the operator defined by an acyclic connection over Y. There are a number of equivalent definitions of Sobolev spaces L^p_k, just as we have seen already in Subsection 3.1.1 for the case $p = 2$. We will take, as a starting point, the definition of the L^p_k norm (initially on compactly supported, smooth, sections) through the formula

$$\|f\|_{L^p_k} = \left(\sum_{j=0}^{k} \int_{Y \times \mathbf{R}} |\nabla^{(j)} f|^p \right)^{1/p}$$

where $\nabla^{(j)}$ is the j-fold iterated covariant derivative. We could define another norm $\| \ \|'_{L^p_k}$ inductively using the operator D:

$$\|f\|'_{L^p_k} = \|Df\|'_{L^p_{k-1}} + \|f\|_{L^p}.$$

To see that these are equivalent norms, and hence define the same completion L^p_k, we divide the tube into a set of bands B_n, the integer translates of the model band $B = Y \times (0, 1)$. We let B_n^+ be the translates of a slightly larger band, so we can apply Inequalities 3.5 and 3.7 to the pairs (B_n, B_n^+), with constants which are plainly independent of n. Then, when $k = 1$, we have

$$\|\nabla f\|^p_{L^p} = \sum_{n=-\infty}^{\infty} \int_{B_n} |\nabla f|^p \leq \text{const.} \sum_{n=-\infty}^{\infty} \left(\int_{B_n^+} |Df|^p + \int_{B_n} |f|^p \right) \quad (3.8)$$

$$\leq 2 \text{ const.} \|Df\|^p_{L^p} + \text{ const.} \|f\|^p_{L^p}. \quad (3.9)$$

Here the factor of 2 comes from the overlap between different sets B_n^+. This inequality shows that the norms $\| \ \|_{L^p_1}$, $\| \ \|'_{L^p_1}$ are equivalent, and the argument for larger k is similar. Of course there are many other possible definitions of an L^p_k norm over a manifold with tubular ends, which turn out to be equivalent.

We can obtain Sobolev inequalities in a similar fashion, appealing to the results over bands. Let p, q be indices related by Equation 3.6 (with

$d = 4$ of course). Then $q > p$ and so for any positive a_n,

$$\sum_{n=-\infty}^{\infty} a_n^{q/p} \leq \left(\sum_{n=-\infty}^{\infty} a_n \right)^{q/p}.$$

Now, for any f on $Y \times \mathbf{R}$:

$$\begin{aligned} \|f\|_{L^q}^q &= \sum_n \int_{B_n} |f|^q \\ &\leq \text{const.} \sum_n \left(\int_{B_n} |\nabla f|^p + |f|^p \right)^{q/p} \\ &\leq \text{const.} \left(\sum_n \int_{B_n} |\nabla f|^p + |f|^p \right)^{q/p} \leq \text{const.} \|f\|_{L_1^p}. \end{aligned}$$

In general we have

Proposition 3.20 *If X is a 4-manifold with tubular ends, $k \geq l, q \geq p$ and the indices p, q are related by*

$$k - 4/p \geq l - 4/q$$

then there is a constant

$$c_{X,p,q}$$

such that for any section s of a unitary bundle over X,

$$\|s\|_{L_l^q} \leq c_{X,p,q} \|s\|_{L_k^p}.$$

Let us pause to make three remarks related to the discussion above. First, the results apply very generally. If X is any manifold of *bounded geometry* we can obtain a Sobolev inequality like Proposition 3.20 over X by taking a suitably 'uniform' cover and applying the local result on each member of the cover. Similarly for the equivalent definitions of L_k^p. Second, we should point out that the inequality

$$\|f\|_{L^q} \leq \text{const.} \|f\|_{L_1^p},$$

for compactly supported f, does *not* hold over a tube. To see this one just considers functions with larger and larger supports. The inequality *does* hold over $\mathbf{R}^n$, for suitable exponents, and is indeed what is normally called the Sobolev inequality. This is an instance of the difference between the function theory on Euclidean spaces and on manifolds with tubular ends. The final point to mention is that we also have the pointwise inequality over a tube

$$\|f\|_{L^\infty} \leq \text{const.}\,\|f\|_{L^p_k}, \tag{3.10}$$

when $k - d/p > 0$. This follows immediately by reducing to bands. Thus we get embeddings $L^p_k \subset C^0$.

We now turn to the global properties of the elliptic operator D on L^p spaces. Plainly the operator extends to give a bounded map

$$D : L^p_k \to L^p_{k-1}.$$

The main result we need is an extension of Proposition 3.4:

Proposition 3.21 *The operator D defines a Banach space isomorphism from L^p_k to L^p_{k-1}.*

This will follow easily from a simpler statement. Let Q be the inversion operator constructed in Proposition 3.4, defined initially on the smooth, compactly supported, sections.

Lemma 3.22 *There is a constant C_p such that, for all ρ,*

$$\|Q\rho\|_{L^p} \leq \|\rho\|_{L^p}$$

To deduce Proposition 3.21 from Lemma 3.22 we use the second norm $\|\;\|'_{L^p_k}$ to define our L^p_k spaces: then for any smooth, compactly supported, ρ,

$$\begin{aligned} \|Q\rho\|'_{L^p_k} &= \|DQ\rho\|'_{L^p_{k-1}} + \|Q\rho\|_{L^p} = \|\rho\|'_{L^p_{k-1}} + \|Q\rho\|_{L^p} \\ &\leq \|\rho\|'_{L^p_{k-1}} + \text{const.}\,\|\rho\|_{L^p} \leq \text{const.}\,\|\rho\|'_{L^p_{k-1}}. \end{aligned}$$

So we can extend the operator to an inverse on the completion, as asserted by Proposition 3.21.

To prove Lemma 3.22 we assume to begin with that $p \geq 2$. We will think of the inverse Q as being defined by convolution with an operator-valued function K on $\mathbf{R}$:

$$Q(\rho)_s = \int_{-\infty}^{\infty} K(s-t)\rho_t \, dt.$$

The eigenspace representation of K_τ shows that its $L^2(Y)$-operator norm decays exponentially:

$$\|K(\tau)\| \leq e^{-\delta\tau}.$$

We now wish to appeal to the elementary fact that convolution defines a bounded bilinear map from $L^p \times L^1$ to L^p:

$$\|k * g\|_{L^p(\mathbf{R})} \leq \|k\|_{L^p(\mathbf{R})} \, \|g\|_{L^1(\mathbf{R})}. \tag{3.11}$$

To apply this we introduce, temporarily, a 'mixed' norm $\| \ \|_{p,2}$ over $Y \times \mathbf{R}$

$$\|f\|_{p,2} = \left(\int_{-\infty}^{\infty} \left(\int_Y |f|^2 \right)^{p/2} \right)^{1/p}.$$

Since its operator norm decays exponentially, the kernel K is integrable, as a function with values in the space of operators from L^2 to L^2, so we can apply Formula 3.11, in a version in which g takes values in a Hilbert space and k takes values in the operators on this Hilbert space. We deduce that Q is bounded with respect to the $(p,2)$ norm:

$$\|Q\rho\|_{p,2} \leq \delta^{-1} \|\rho\|_{p,2}. \tag{3.12}$$

We now compare the L^p norm with $\| \ \|_{p,2}$. Suppose for simplicity that Y has volume 1. Then, comparing the L^p norm with the L^2 norm over Y we have, since $p \geq 2$,

$$\|f\|_{p,2} \leq \|f\|_{L^p}, \tag{3.13}$$

so $\|Q\rho\|_{p,2} \leq \|\rho\|_L^p$. Now let $f = Q\rho$, so $Df = \rho$. We consider the situation over one of the bands B_n. The inequalities above give

$$\int_{B_n} |f|^p \leq \text{const.} \left(\int_{B^+(n)} |\rho|^p + \left(\int_{B_n} |f|^2 \right)^{p/2} \right).$$

For a function supported on the finite band B_n the $(p,2)$ norm dominates the L^2 norm. Applying this to the last term in the expression above, then summing over n and writing $f = Q\rho$, we get

$$\int_{Y \times \mathbf{R}} |Q\rho|^p \leq 2 \, \text{const.} \int_{Y \times \mathbf{R}} |\rho|^p + \|Q\rho\|_{p,2}^p.$$

Combined with Formula 3.13 this proves the Lemma, in the case when $p \geq 2$. If $p \leq 2$ we use a standard duality argument. Let p' be the conjugate index to p so that $L^{p'}$ is the dual of L^p. Let Q^* be the adjoint of Q, the inversion operator for the formal adjoint D^*. We can apply the argument above with Q^* in place of Q, so Q^* is a bounded operator on $L^{p'}$. Now, if $\langle \, , \rangle$ denotes the standard pairing between L^p and $L^{p'}$,

$$\|P\rho\|_{L^p} = \sup \langle P\rho, g \rangle,$$

where g runs over the unit ball in $L^{p'}$. Thus

$$\begin{aligned}\sup\langle P\rho, g\rangle &= \sup\langle \rho, P^*g\rangle \leq \sup \|\rho\|_{L^p}\|P^*g\|_{L^{p'}} \\ &\leq \text{const.}\sup\|\rho\|_{L^p}\|g\|_{L^{p'}} \leq \text{const.}\,\|\rho\|_{L^p}.\end{aligned}$$

This completes the proof of Lemma 3.22.

We could now go on to set up the whole linear theory in the L^p setting. In the case when the limits are acyclic the operator D defines a Fredholm map from L^p_k to L^p_{k-1}. The index of this map is independent of p and k, as the reader will easily be able to verify, so is given by the invariant $\mathrm{ind}(\mathbf{P})$ we studied above. We can also introduce exponentially weighted norms,

$$\|f\|_{L^{p,\underline{\alpha}}_k} = \|Wf\|_{L^p_k},$$

and obtain Fredholm operators with indices $\mathrm{ind}^+, \mathrm{ind}^-$. If $\underline{\alpha}$ is a vector of weights α_i, one for each end of a 4-manifold X, we let $L^{p,\underline{\alpha}}_k$ be the norm obtained from a weight function W as before. We have then

Proposition 3.23 *Suppose each weight α_i is positive. Then for $k > l$, $k - 4/p = l - 4/q$,*

$$L^{p,\underline{\alpha}}_k \subset L^{q,\underline{\alpha}}_l,$$

and for $k - 4/p > 0$

$$L^{p,\underline{\alpha}}_k \subset C^0.$$

This is a trivial consequence of the corresponding unweighted inequality, replacing f by Wf.

4

Gauge theory and tubular ends

This Chapter occupies a central position in the book as a whole. Building on the results of Chapter 3, we develop the theory of instantons over non-compact manifold with tubular ends. The theory can be considered as a modification of the standard set-up over compact manifolds and we shall see that many of the results go over to the non-compact case; in particular the instanton equations become non-linear Fredholm equations and we get finite-dimensional moduli spaces of solutions. The original reference for most of this material is [45] (in the more general setting of manifolds with 'periodic ends').

The work in this Chapter falls into three main parts. First we study the decay of instantons over tubular ends. We shall see that, under suitable conditions, an instanton with L^2 curvature can be represented by a connection form which decays exponentially down the tube. This can be regarded as a counterpart of the elliptic regularity theory for the instanton equations; it implies that all of the possible natural definitions of 'decaying instantons' are equivalent. The proofs are straightforward modifications of those of previous results for the case when the cross-section is a 3-sphere. In the second part we set up function spaces to present the instanton equations as non-linear Fredholm equations. We shall consider two cases, depending upon the form of the limiting connections over the cross-sections of the tubes. The simplest case is when these are *acyclic*; recall that this means that they are *irreducible* and the twisted cohomology groups are all zero. The slightly more difficult case which we shall treat in detail arises when the limiting connection is non-degenerate but reducible, in the terminology of Chapter 2. For example, this occurs if the limit is the trivial connection over a homology 3-sphere. In this case one has to introduce the weighted spaces to obtain a Fredholm theory, and these bring in some new features.

The third part of the work in this Chapter treats the 'gluing theory' for instantons over manifolds with tubular ends. Again, we treat the acyclic case first before moving on to the more complicated situation where the limits are reducible. Finally, in the Appendix to this Chapter, we discuss other extensions of the theory in which we allow non-isolated limits, and give a few additional analytical results (which will not be used in the rest of the book).

4.1 Exponential decay

We return to the now familiar picture of a 4-manifold X with tubular ends and assume that for each end $Y \times (0,\infty) \subset X$ the cross-section Y satisfies the *non-degeneracy* condition of Chapter 2, i.e. for each flat $SU(2)$ connection ρ over Y the cohomology group $H^1(Y,\rho)$ is zero.

Now let A be an instanton over X and suppose the curvature $F(A)$ is in L^p for some $p \geq 2$. (Recall that the L^p norm gives *weaker* control of decay behaviour for larger p.) The first result we have is

Proposition 4.1 *Under these conditions, for each end $Y \times (0,\infty)$ of X there is a flat connection ρ over Y (unique up to equivalence) such that A converges to ρ, in the sense that the restrictions $A|_{Y\times\{T\}}$ converge (modulo gauge equivalence) in C^∞ over Y as $T \to \infty$.*

We consider a family of bands $B_T = Y \times (T-1, T)$ which we identify with the model $B = Y \times (0,1)$ by translation. Let A_T be the connection over B obtained from the restriction of A to B_T in this way, so the integrability of $|F(A)|^p$ over the end implies that

$$\|F(A_T)\|_{L^p(B)} \to 0 \quad \text{as} \quad T \to \infty.$$

Uhlenbeck's weak compactness result from Chapter 2 now implies that for any sequence $T_i \to \infty$ there are a subsequence T_i' and a flat connection ρ over B such that, after suitable gauge transformations,

$$A_{T_i'} \to \rho$$

in C^∞ over compact subsets of B. In particular the restriction of $[A_{T_i'}]$ to the cross-section $Y \times \{1/2\}$ converges in C^∞ to ρ.

It is now easy to see that, since the flat connections over Y are isolated, the limit is unique – independent of the sequence and subsequence chosen. First consider the metric $d(\,,)$ on the space of equivalence classes $\mathcal{B}_Y$ of connections over Y induced from the L^2 metric over Y. We have shown that the continuous path $[A_T]$ in $\mathcal{B}_Y$ converges to the space of flat connections $\mathcal{R} \subset \mathcal{B}_Y$, in the sense that $d([A_T], \mathcal{R}) \to 0$. But the points of $\mathcal{R}$ are isolated so there is a $\delta > 0$ with $d(\rho, \sigma) \geq \delta$ for distinct points ρ, σ in $\mathcal{R}$. For large T the connections $[A_T]$ have distance less than $\delta/2$ from some point of $\mathcal{R}$. But by the intermediate value theorem applied to the continuous functions $d(\rho, [A_T])$ this point must be independent of T, so we obtain a flat connection ρ such that A_T converges to ρ in the L^2 distance. But now it follows similarly that the convergence is in C^∞, since any sequence has a C^∞-convergent subsequence.

The effect of this result is that we can partition these L^p-integrable instantons into classes labelled by the limiting flat connection over the ends. We shall now go on to study this limiting behaviour in more detail. We may as well fix attention on a connection A over a half-tube $Y \times (0, \infty)$ which converges in the sense above to a given flat connection ρ at infinity, where $H^1(Y; \mathrm{ad}\,\rho) = 0$. Let δ be the smallest positive eigenvalue of L_ρ restricted to $\ker d_\rho^* \subset \Omega^1$.

Theorem 4.2 *If $F(A)$ is in L^p for some $p \geq 2$ then there is a constant C such that*

$$|F(A)| \leq C e^{-\delta t}.$$

We prove this using a differential inequality derived from the instanton equation. For clarity we first suppose that we can take $p = 2$, and for $T > 0$ we set

$$J(T) = \int_T^\infty |F(A)|^2.$$

Now we can on the one hand express $J(T)$ as the integral of the Chern–Weil 4-form $-\mathrm{Tr}(F^2)$ over $Y \times (T, \infty)$, using the basic property 2.15. By taking the limit over finite tubes $Y \times (T, T')$ with $T' \to \infty$ we see that

$$(8\pi^2)^{-1} J(T) = \vartheta(A_T) - \vartheta(\rho), \tag{4.1}$$

where ϑ is the Chern–Simons function, and as above A_T is the connection over Y obtained by restriction to $Y \times T$. Since A_T is close to ρ we can choose a local lifting of the Chern–Simons function to $\mathbf{R}$, so Equation 4.1 holds in $\mathbf{R}$ rather than just $\mathbf{R}/\mathbf{Z}$.

On the other hand the T derivative of J can obviously be expressed as minus the integral over $Y \times \{T\}$ of the curvature density $|F(A)|^2$, and this is exactly twice the 3-dimensional curvature density $2|F(A_T)|^2$, by the relation Equation 2.10 between the two components of the curvature for an instanton over a tube. Thus

$$\frac{dJ}{dT} = -2\|F(A_T)\|^2_{L^2(Y)}. \tag{4.2}$$

To connect these two observations we establish an inequality between the gauge-invariant quantities $\vartheta(A_T)$ and $\|F(A_T)\|_{L^2}$, valid for any connection over Y which is close to ρ. We write, for fixed large T,

$$A_T = \rho + a,$$

so we may suppose that a is as small as we please in C^∞. Also, we may suppose that a satisfies the Coulomb condition:

$$d_\rho^* a = 0.$$

Now, using Equation 2.18,

$$\vartheta(A_T) - \vartheta(\rho) = \int_Y \mathrm{Tr}\left(a \wedge d_\rho a + \frac{2}{3} a \wedge a \wedge a \right). \tag{4.3}$$

On the other hand $F(\rho + a) = d_\rho a + a \wedge a$ so

$$\|F_{A_T}\|^2_{L^2(Y)} \leq \|d_\rho a\|^2 + \|a \wedge a\|^2 + 2|\langle d_\rho a, a \wedge a\rangle|. \tag{4.4}$$

Consider now the quadratic terms in Formulae 4.3 and 4.4. The Coulomb condition $d_\rho^* a = 0$ means that $*d_\rho a = L_\rho a$ and then, from the spectral decomposition for L and the definition of δ, we see that

$$\int_Y \mathrm{Tr}(a \wedge d_\rho a) = \langle a, L_\rho a\rangle \leq \delta \|L_\rho a\|^2 = \delta \|d_\rho a\|^2.$$

We next estimate the remaining terms in the two expressions. We use the fact that the kernel of L_ρ in Ω^1 is trivial, so

$$\|a\|_{L^2_1} \leq \text{const.}\ \|L_\rho a\|_{L^2},$$

and combine this with the Sobolev embeddings $L^2_1 \to L^r$, for $r \leq 6$, in three dimensions. We get

$$\vartheta(A_T) - \vartheta(\rho) \leq \delta \|d_\rho a\|^2 + \text{const.}\ \|d_\rho a\|^3,$$

and

$$\|d_\rho a\|^2 - \|F(A_T)\|^2 \leq \text{const.}\ (\|d_\rho a\|^3 + \|d_\rho a\|^4).$$

(Here all norms are L^2.)

This gives, when $d_\rho a$ is small enough (i.e. when T is large enough),

$$\vartheta(A_T) - \vartheta(\rho) \leq \delta \|F(A_T)\|^2 + \text{const.}\ \|F(A_T)\|^3. \tag{4.5}$$

Putting all this together, we get a differential inequality

$$\frac{dJ}{dt} \leq -\delta J + \text{const.}\, J^{3/2}.$$

It is easy to see that this implies that J decays exponentially. Indeed, for any $\delta' \leq \delta$ we have $dJ/dt \leq -\delta' J$, for $T \geq T_0$ say, and this inequality immediately gives $J(T) \leq Ce^{-\delta' T}$, with $C = J(T_0)e^{\delta' T_0}$. From this one can go back to improve the exponent to δ, although this is not really important for us.

Finally, having obtained the exponential decay of J we deduce that of the curvature density itself via elliptic estimates on the model band. Notice that we can get get estimates on all the covariant derivatives of curvature in this way. Obviously $J(T)$ dominates the L^2 norm of the curvature of A over the band B_T, and this gives a bound on all higher derivatives over an interior domain.

We now go back to consider the case when $F(A)$ is only assumed to lie in L^p for some $p > 2$. We introduce a small positive parameter σ and set now

$$J(T) = \int_T^\infty e^{-\sigma t} |F(A)|^2.$$

The integral converges because of the exponential factor. Now

$$dJ/dT = e^{-\sigma T} \|F(A_T)\|^2.$$

On the other hand, integrating by parts, we get

$$J(T) = e^{-\sigma T}(\vartheta(A_T) - c) + \sigma \int_T^\infty e^{-\sigma t}(\vartheta(A_t) - c),$$

for any constant c. We can choose $c = \vartheta(\rho)$ so, applying the inequality Formula 4.5 to each connection A_t for $t \in [T, \infty)$, we have for large T

$$J(T) \leq 2\delta^{-1}\left(-\frac{dJ}{dT}\right) + 2\sigma\delta^{-1} J(T).$$

Now if we choose $\sigma < \delta/2$, say $\sigma = \delta/4$, we can rearrange this to get

$$\frac{dJ}{dT} \leq -4\delta^{-1} J(T),$$

which implies that J decays exponentially, so the curvature is in L^2 after all.

Now write A_0 for the flat connection on a bundle P_0 over the half-tube lifted from ρ over Y.

Proposition 4.3 *Let A be a connection on a bundle Q over X which satisfies the conditions stated at the beginning of this Section. For each end $Y \times (0, \infty) \subset X$ there is a bundle map $\chi : P_0 \to Q|_{Y \times (0,\infty)}$ such that $A = \chi^*(A_0) + a$ with*

$$|a| \leq \text{const.}\, e^{-\delta t}.$$

Moreover, we can choose a so that all derivatives decay exponentially:

$$|\nabla^{(l)} a| \leq \text{const.}\, e^{-\delta t}.$$

Choose first any temporal gauge representation $A_0 + b$; so b has zero dt component as in Chapter 2. The instanton equation $db/dt = *_3 F$ gives

$$\left| \frac{da}{dt} \right| = |F(A)_t| \leq \text{const.}\, e^{-\delta t},$$

so b_t converges to some limit b_∞, at an exponential rate:

$$|b_t - b_\infty| \leq \text{const.}\, e^{-\delta t}.$$

Similarly, the decay of the derivatives of the curvature of A on the tube implies that b_∞ is smooth and all Y derivatives of b_t converge. Now we know that the gauge equivalence classes $[\rho + b_t]$ converge to $[\rho]$, i.e. we can find gauge transformations g_t over Y so that $g_t(\rho + b_t) \to \rho$. After possibly taking a subsequence we can suppose then that the g_t converge in C^∞ to some limit g and $g(\rho + b_\infty) = \rho$ (compare [17][Section 2.3.7]). Then we apply the constant gauge transformation g over the tube to modify the representative to $a_t = b_t + g(\rho) - \rho$, which tends to zero at infinity. Replace a by $\rho + a - g(\rho)$ to get a representative which tends to zero at infinity.

We shall also need a variant of these decay results for connections over long, but finite, tubes, in the same spirit as Lemma 3.2. Recall that we denote by $B_T^{\pm}$ the bands $Y \times \pm(T-1, T) \subset Y \times \mathbf{R}$.

Proposition 4.4 *Suppose all flat connections over Y are non-degenerate. There are constants C, δ, C_l such that for any $T > 2$ and any instanton A over $Y \times (-T, T)$ with $\|F(A)\|^2 < C$ there is a flat connection A_0 over the tube such that A is gauge-equivalent to $A_0 + a$ with*

$$|\nabla^{(l)} a| \leq C_l e^{-\delta(T - |t|)}(I_+ + I_-),$$

over $Y \times (-T + \frac{1}{2}, T - \frac{1}{2})$, *where*

$$I_{\pm}^2 = \int_{B_T^{\pm}} |F(A)|^2.$$

To prove this one follows much the same argument as before, deriving a differential inequality for

$$I(t) = \int_{Y \times (-t,t)} |F(A)|^2.$$

4.2 Moduli theory

The decay results of the previous Section mean there is little possibility of argument about the appropriate moduli spaces of instantons over the manifold X, at least in the case when the flat connections over the ends are non-degenerate. The union of the moduli spaces, as a set, will be the gauge equivalence classes of instantons with L^p curvature and we have seen that this is independent of the choice of exponent $p \geq 2$. (More generally, the arguments of the previous Section show that any sensible definition of instantons which are 'flat at infinity' will agree with this one.) Moreover we have seen that we can assign to any such instanton A a definite limit over each end of X and hence an adapted bundle P. Thus we obtain moduli spaces M_P, labelled by the adapted bundles.

In this Section and the next we will study one of these moduli spaces M_P (which we will normally denote by M). We wish to describe it in the same general framework which we are familiar with in the compact case, and for this we need to choose suitable function spaces. Of course, as in the compact case, there is a good deal of choice in these – one can use many different function spaces which give the same moduli space. We will treat two cases; first the 'acyclic case', which is rather straightforward, and then, in Section 4.3, the non-degenerate case. For the remainder of Section 4.2 then we suppose that P is an adapted bundle whose limits over the ends of X are all acyclic. We let A_0 be any connection on P which agrees with the flat connections over the ends, under the fixed identifications given by the definition of an adapted bundle. We will write ∇_0 for the covariant derivative of A_0, and similarly for other operators. We fix an exponent p with $2 < p < 4$ and let $q = \frac{4p}{(4-p)}$, so $q > 4$. Note also that $q > 2p$.

We define our space of connections $\mathcal{A}$ to be those of the form $A_0 + a$ with a in L_1^p, i.e.

$$|a|, \ |\nabla_0 a| \in L^p.$$

The curvature of $A_0 + a$ differs from that of A_0 by $d_0 a + a \wedge a$. The linear term is obviously in L^p and the same is true of the quadratic term by the Sobolev theorem (Proposition 3.20) which tells us that

$$a \in L^p_1 \Rightarrow a \in L^q.$$

So $|a|$ is in $L^p \cap L^q$, and since $p < 2p < q$ we obtain by Hölder's inequality that $|a| \in L^{2p}$. Hence $a \wedge a$ is indeed in L^p. Now taking the self-dual component of the curvature we get a map

$$F^+ : \mathcal{A} \to L^p(\Omega^+(\mathfrak{g}_P))$$

and it is clear that this is a smooth map. (In this Section we shall normally emphasise in our notation the function spaces rather than the differential-geometric aspects, so we will for example write $F^+ : \mathcal{A} \to L^p$.)

Similarly, we can for each A in $\mathcal{A}$ regard the operator $D_A = D_0 + a$ as a bounded map

$$D_A : L^s_1 \to L^s$$

for any s with $1 < s < 4$. To see this we use the embedding

$$L^s_1 \subset L^s \cap L^r,$$

with $r = 4s/(4-s)$; then $q > 4$ implies that

$$\frac{1}{r} + \frac{1}{q} \le \frac{1}{s}.$$

On the other hand $\frac{1}{s} \le \frac{1}{s} + \frac{1}{p}$, so Hölder's inequality tells us that multiplication maps

$$(L^s \cap L^r) \times (L^p \cap L^q) \to L^s, \tag{4.6}$$

and we can apply this to the algebraic action of a on L^s_1.

An argument of just the same kind as that in the discussion following Corollary 3.7 shows that D_A is Fredholm on L^s_1 and that the index is independent of $A \in \mathcal{A}$. Thus to compute the index we can reduce to the model case D_0 which was treated in Chapter 3, and we get

Proposition 4.5 *For any A in $\mathcal{A}$ the index of $D_A : L^s_1 \to L^s$ is given by the invariant* ind P.

We now turn to gauge transformations. The appropriate definition of equivalence of connections in $\mathcal{A}$ is immediate – we say A_1, A_2 are equivalent if there is an $L^p_{2,\mathrm{loc}}$ bundle automorphism g such that $g(A_1) = A_2$.

Notice, of course, that we are in the range where $L^p_{1,\mathrm{loc}}$ functions are *continuous*, so there is no difficulty at all in interpreting what one means by a gauge transformation of this class. Our task is to show that this equivalence relation is generated by the action of an appropriate gauge group $\mathcal{G}$ on $\mathcal{A}$. The definition of $\mathcal{G}$ is forced on us. If g is an $L^p_{2,\mathrm{loc}}$ gauge transformation then $g(A_0)$ is in $\mathcal{A}$ if and only if $d_0 g g^{-1}$ is in L^p_1, so we put

$$\mathcal{G} = \{g \in \mathrm{Aut}\, P : d_0 g g^{-1} \in L^p_1\}. \tag{4.7}$$

Now we know that $L^p_1 \subset L^q$ and so, since also g is pointwise bounded, $d_0 g$ is in $L^p \cap L^q$ for $g \in \mathcal{G}$. In particular $d_0 g$ is in L^4, since $p < 4 < q$. Now Hölder's inequality tells us that multiplication is a bounded map:

$$L^4 \times L^q \to L^p.$$

Let us write $\alpha = d_0 g g^{-1}$, so $d_0 g = \alpha g$ and $\nabla_0(d_0 g) = (\nabla_0 \alpha) g + \alpha(\nabla_0 g)$. The multiplication Formula 4.6 implies that the second term is in L^p, and the first is obviously so since g is bounded. Thus $d_0 g$ is in L^p_1 and by just the same argument we see that

$$\mathcal{G} = \{g \in \mathrm{Aut}\, P : \nabla_0 \nabla_0 g, \nabla_0 g \in L^p\}.$$

Likewise, if $g, h \in \mathcal{G}$ then

$$\nabla_0 \nabla_0 (gh) = (\nabla_0 \nabla_0 g) h + 2 \nabla_0 g \nabla_0 h + g \nabla_0 \nabla_0 h,$$

and the multiplication property shows this is in L^p. It follows that (pointwise) multiplication is defined in $\mathcal{G}$. Also, expanding out the derivatives of g^{-1}, we see that this is in $\mathcal{G}$, so $\mathcal{G}$ is a group Similarly, if $A_0 + a \in \mathcal{A}$ and $g \in \mathcal{G}$ then

$$g(A_0 + a) = A_0 - (d_0 g) g^{-1} + g a g^{-1},$$

and using the same multiplication property we can see that gag^{-1} is in L^p_1, so $g(A_0 + a) \in \mathcal{A}$ and $\mathcal{G}$ acts on $\mathcal{A}$. Conversely, the same sort of argument shows that if A_1, A_2 are any two elements $\mathcal{G}$ and if g is an $L^p_{2,\mathrm{loc}}$ gauge transformation with $g(A_1) = A_2$ then g is in $\mathcal{G}$. Thus $\mathcal{G}$ does generate the required equivalence relation.

Next, we can define a topology on this gauge group $\mathcal{G}$. We fix a base point x_0 in X and define a system of neighbourhoods of the identity

$$U_\epsilon = \{g \in \mathcal{G} : \|\nabla_0 \nabla_0 g\| \ \leq \epsilon, \ \|\nabla_0 g\| \leq \epsilon, |g(x_0) - 1| \leq \epsilon\}. \tag{4.8}$$

These define a topology which is independent of the base point x_0 in X (and so it is finer than the topology of uniform convergence over compact subsets). Moreover it is easy to check that the action $\mathcal{G} \times \mathcal{A} \to \mathcal{A}$ (with the L^p_1 topology on $\mathcal{A}$) is continuous.

All of this is quite satisfactory, but there is one further fact to establish in order to work effectively with this gauge group. We want to show that the L^p norms of the first two derivatives of g control the deviation of g from the identity. To do this we have to use the hypothesis that the limiting connections are acyclic. Note that this hypothesis has *not* been used in the arguments above. We state the results in the next two Propositions.

Proposition 4.6 *If the limiting connections ρ_i are irreducible then there is a constant C such that for any section ξ of $\mathfrak{g}_P$ with $\nabla_0 \xi$ and $\nabla_0 \nabla_0 \xi \in L^p$ we have $|\xi| \to 0$ at infinity in X and*

$$\sup |\xi| \leq C\ (\|\nabla_0 \nabla_0 \xi\|_{L^p} + \|\nabla_0 \xi\|_{L^p}).$$

The proof is a minor modification of that for Formula 3.10 (which involved also the L^p norm of the section itself). If the covariant derivative ∇_{A_0} has trivial kernel, i.e. if the connection ρ is irreducible, then over the the model band $B = (0,1) \times Y$ we have an inequality

$$\|f\|_{C^0(B)} \leq \|\nabla_0 f\|_{L^q(B)}$$

and we can apply this inequality, over the translates of B, to the restrictions of ξ. We will write $L(\mathcal{G})$ for the set of sections of $\mathfrak{g}_P$ considered in Proposition 4.6.

Next we have a corresponding result for gauge transformations:

Proposition 4.7 *Under the conditions of Proposition 4.6, if g is in $\mathcal{G}$ then either $|g(x) - 1| \to 0$ or $|g(x) + 1| \to 0$ as x tends to infinity in X.*

Again we consider the picture over the model band. If g_n is a sequence of gauge transformations over B with $\|\nabla_0 \nabla_0 g_n\|_{L^p}$, $\|\nabla_0 g\|_{L^p} \to 0$ then by the Rellich lemma there is a uniformly convergent subsequence, and the limit g_∞ must satisfy $\nabla_0 g_\infty = 0$. By our hypothesis this implies that $g_\infty = \pm 1$. We apply this to the restriction of a gauge transformation over the end of X to the translates of B: arguing just as in the proof of Proposition 4.1 we see that g must converge to either plus or minus 1 at each end.

By the same argument we see that, in the acyclic case, the topology on $\mathcal{G}$ is stronger than the uniform topology, that is:

Proposition 4.8 *If* $g_n \to 1$ *in* $\mathcal{G}$ *then* $\sup_X |g_n(x) - 1| \to 0$.

This has the important consequence that *in the acyclic case* we can effectively linearise about the identity.

Proposition 4.9 (i) *The pointwise exponential map* $\exp : L(\mathcal{G}) \to \mathcal{G}$ *defines a homeomorphism from sufficiently small neighbourhoods of the origin in* $L(\mathcal{G})$ *to neighbourhoods of the identity in* $\mathcal{G}$, *and the translates of these neighbourhoods make* $\mathcal{G}$ *into a Banach Lie group modelled on* $L(\mathcal{G})$.

(ii) *The action of* $\mathcal{G}$ *on* $\mathcal{A}$ *is smooth.*

The proofs are routine, for example in (i) one merely has to use the fact that if ξ is small the derivatives of ξ and $\exp(\xi)$ are comparable.

We are now in a good position to study the quotient $\mathcal{B}$ of $\mathcal{A}$ by $\mathcal{G}$, in the acyclic case. The extra ingredient we need is a system of slices for the action, and these are provided by the usual Coulomb gauge condition. By the discussion in the last part of Chapter 3 we have a direct sum decomposition

$$L_1^p = (L_1^p \cap \ker d_0^*) \oplus d_0(L_2^p).$$

The same argument shows that the corresponding decomposition holds for any connection A in $\mathcal{A}$:

$$L_1^p = (L_1^p \cap \ker d_A^*) \oplus d_A(L_2^p).$$

Thus the implicit function theorem can be applied to show that the Coulomb slices

$$T_{A,\epsilon} = \{A + a : d_A^* a = 0,\ \|a\|_{L_2^p} \leq \epsilon\}$$

give local transversals for the action. Notice that there can be no reducible connections in $\mathcal{A}$, since the connections are assumed to be irreducible at infinity.

We obtain then

Proposition 4.10 *The quotient* $\mathcal{B}$ *is a Banach manifold modelled on the* $T_{A,\epsilon}$ *and the self-dual part of the curvature defines a section of a bundle over* $\mathcal{B}$ *which is Fredholm, with index* $\operatorname{ind} P$.

The significance of the index is that it gives the 'virtual dimension' of the instanton moduli space. If a point $[A]$ in the moduli space is 'regular' in that

$$d_A^+ : L_1^p \to L^p$$

is surjective (or equivalently D_A is surjective), then a neighbourhood of $[A]$ in the moduli space is a manifold of this dimension. More generally, the virtual dimension gives the dimension of the moduli space of solutions of a generic perturbation of the equation. In fact there is no difficulty in extending the discussion for compact manifolds to show that, so long as the instantons are not flat, we can make these perturbations by changing the Riemannian metric on X. In fact the argument of Freed and Uhlenbeck, as presented in [17], shows that it suffices to make an arbitrarily small change in the metric over an arbitrarily small neighbourhood in X (so in particular we do not need to modify the metric over the ends). Recall that the instantons on an adapted bundle P are flat if and only if the relative Chern class $\kappa(P) \in \mathbf{R}$ is zero. To sum up, we obtain

Theorem 4.11 *For generic metrics on X all non-flat instanton solutions are regular, and all moduli spaces M_P with $\kappa(P) \neq 0$ are smooth manifolds of dimension* $\dim M_P = \mathrm{ind}(P)$.

4.3 Moduli theory and weighted spaces

We now wish to consider how the moduli theory of the previous Section must be changed in the case when some of the limiting flat connections are reducible. The first point in the previous Section where the irreduciblity hypothesis was required was in Proposition 4.6 – the fact that the sections of $\mathfrak{g}_P$ whose first two derivatives lie in L^p tend to 0 at each end of X. If the limiting connections are reducible this is no longer true. The prototype case is when the connection A_0 is the product connection on the trivial bundle, so ∇_0 is just ordinary differentiation of functions. Consider for example a function ξ which equals t^γ on an end. Then $\nabla\xi$ and $\nabla^2\xi$ are in L^p if $\gamma \leq 1 - 1/p$, but if $\gamma > 0$ the function is unbounded. Thus we cannot invert the pointwise exponential map on any neighbourhood U_ϵ of the identity and $\mathcal{G}$, as defined above, is not a Lie group. (So, for example, we cannot apply the implicit function theorem to linearise the action of $\mathcal{G}$ on the L_1^p connections.)

To get around this problem we will introduce weights into our function spaces, but before embarking on this we will examine the breakdown in the set-up above in more detail. Let ρ be a reducible flat connection over one of the ends $Y \times (0, \infty)$, let $K_\rho \subset \Omega^0_Y(\mathfrak{g}_P)$ be the space of covariant-constant sections, and $K_\rho^\perp \subset \Omega^0_Y(\mathfrak{g}_P)$ be the L^2 orthogonal complement. Then if ξ is a section of $\mathfrak{g}_P$ over the half-tube we can write $\xi = k + \xi'$ by decomposing for each time t

$$\xi_t = k_t + \xi'_t \in K_\rho \oplus K_\rho^\perp. \tag{4.9}$$

Then the derivative of k in the Y direction is zero by construction, and the derivative in the t variable can be obtained by projecting $\nabla_0 \xi$ to the subspace K_ρ. So the L^p norms of the t derivatives of ξ control those of k. Hence $\nabla_0^2 \xi,\ \nabla_0 \xi < \infty$ implies $\nabla_0^2 \xi',\ \nabla_0 \xi',\ \nabla_0^2 k,\ \nabla_0 k < \infty$. Now the argument we used in the previous case applies to ξ', so we see that $\xi' \to 0$ at infinity. The problem comes about entirely from the finite-dimensional component k, where we only know

$$\frac{dk}{dt}, \frac{d^2 k}{dt^2} \in L^p$$

and, as we have seen, this does not control k over $[0, \infty)$.

Suppose, however, we were given that $\nabla_0 \xi$ was also in L^1 over the tube. Then the same argument will show that, in this decomposition, dk/dt is in L^1; so k, and hence also ξ, tends to a limit at infinity. Similarly, working in the gauge group rather than its Lie algebra, one sees that if g is a bundle automorphism such that $\nabla_0^2 g, \nabla_0 g \in L^p$ and $\nabla_0 g \in L^1$ then $g(x)$ tends to a limit in the isotropy group Γ_ρ as X tends to infinity in X.

To proceed now with the theory: we choose an exponent $\alpha > 0$ as in Chapter 3, less than the first positive eigenvalue of L_ρ, and define weighted spaces of forms $L^{p,\alpha}_k$ as before. We consider the space of connections

$$\mathcal{A}^\alpha = \{A_0 + a : a \in L^{p,\underline{\alpha}}_1\}.$$

Our previous results carry over to this class. These connections have curvature in $L^{p,\alpha}$ since

$$\begin{aligned} \|a \wedge a\|_{L^{p,\underline{\alpha}}} &= \|Wa \wedge a\|_{L^p} \leq \|\sqrt{W} a\|^2_{L^{2p}} \\ &\leq \text{const.}\, \|Wa\|^2_{L^{2p}} \leq \text{const.}\, \|a\|^2_{L^{p,\underline{\alpha}}_1}. \end{aligned} \tag{4.10}$$

Here we have used Proposition 3.23 and the fact that the weight function W is bounded below, so $\sqrt{W} \leq \text{const.}\, W$. Our decay results show that all the instantons we need to study are captured by these spaces.

Following the lines of the previous Section, we define a group $\mathcal{G}^\alpha$ of gauge transformations g such that

$$\nabla_0 g \in L_1^{p,\underline{\alpha}}.$$

Now, by Hölder's inequality,

$$\int_{Y\times[0,\infty)} |\nabla_0 g| \leq \left(\int_{Y\times[0,\infty)} |\nabla_0 g|^p e^{p\alpha t}\right)^{1/p} \left(\int_{Y\times[0,\infty)} e^{-p'\alpha t}\right)^{1/p'},$$

where p' is the conjugate exponent. So, from the discussion above, each element g of $\mathcal{G}^\alpha$ tends to a limit in Γ_ρ at infinity, and we get an evaluation homomorphism

$$ev : \mathcal{G}^\alpha \to \Gamma_\rho. \tag{4.11}$$

Similarly, in the bundle of Lie algebras, if we set

$$L(\mathcal{G}^\alpha) = \{\xi \in \Omega_X^0(\mathfrak{g}_P) : \nabla_0 \xi \in L_1^{p,\underline{\alpha}}\},$$

then the elements of $L(\mathcal{G}^\alpha)$ tend to limits at infinity, and we again have an evaluation map ev, taking $L(\mathcal{G}^\alpha)$ to K_ρ. The evaluation maps are continuous: indeed we can suppose our base point x_0 is on one end of the tube in X, then the evaluation at infinity $ev(\xi)$ is equal to $\lim k(x)$ which is $k(x_0) + \int k'\, dt$, and this varies continuously with ξ by the inequality above.

We now have

Proposition 4.12 *The pointwise exponential map defines charts making the gauge group $\mathcal{G}^\alpha$ into a Banach Lie group modelled on $L(\mathcal{G}^\alpha)$, acting smoothly on $\mathcal{A}^\alpha$. The evaluation map at each end is a smooth Lie homomorphism, with surjective derivative.*

Again, the verification of these properties is straightforward. Note that the kernel of the evaluation maps is a smooth subgroup $\mathcal{G}_0^\alpha \subset \mathcal{G}^\alpha$ say, the analogue of the 'based' gauge group in the usual compact case. By construction

$$\mathcal{G}^\alpha / \mathcal{G}_0^\alpha \cong \Gamma_\rho.$$

It is at this stage that we bring in the discussion of Subsection 3.3.3. Recall that we introduced an operator $d_A^{*,\underline{\alpha}}$ which is the adjoint of the covariant derivative with respect to the weighted norms. The importance of this is that it gives the appropriate 'Coulomb gauge condition' in this setting; i.e. there is a decomposition of bundle-valued 1-forms

$$L_1^{p,\alpha} = \left(\ker d_A^{*,\underline{\alpha}} \subset L_1^{p,\alpha}\right) \oplus d\left(L_2^{p,\alpha}\right).$$

The proof of this follows just the same pattern as in the unweighted case. However, now we must beware, because the Lie algebra $L(\mathcal{G}^\alpha)$ is *not* the same as L_2^p. Sections in the latter space tend to zero at infinity, whereas those in the former tend to constant values in K_ρ. The subspace $L_2^p \subset L(\mathcal{G}^\alpha)$ is precisely the kernel of the evaluation map at infinity, and this is the Lie algebra of the subgroup $\mathcal{G}_0^\alpha$. Thus the Coulomb condition gives us slices for the action of $\mathcal{G}_0^\alpha$. Thus we can construct a quotient $\tilde{\mathcal{B}} = \mathcal{A}^\alpha/\mathcal{G}^\alpha$. The action of $\mathcal{G}^\alpha$ is free and the Coulomb slices make $\tilde{\mathcal{B}}$ into a Banach manifold, with local models

$$T\tilde{\mathcal{B}}_{[A]} = \ker d_A^{*,\alpha} \cap L_1^{p,\alpha}.$$

Of course there is some extra symmetry which has been ignored in forming this quotient. The full quotient $\mathcal{B} = \mathcal{A}^\alpha/\mathcal{G}^\alpha$ is obtained by dividing $\tilde{\mathcal{B}}$ by the natural Γ_ρ-action, and the dense open subset $\mathcal{B}^* \subset \mathcal{B}$ of irreducible connections is a smooth manifold, a $(\Gamma_\rho/\pm 1)$-quotient of a corresponding subset of $\tilde{\mathcal{B}}$.

We now proceed to the instanton equation over X. This defines a Γ_ρ-invariant Fredholm section of a bundle over $\tilde{\mathcal{B}}$ with linearisation given by the operators $D_{A,\underline{\alpha}}$. The virtual dimension of the moduli space $\tilde{M} \subset \tilde{\mathcal{B}}$ is thus the index of this operator in the weighted space, which is $\mathrm{ind}^+(P) + \dim H_\rho^0$, by Proposition 3.19. The genuine moduli space M_P is the quotient $\tilde{M}/\Gamma_\rho$ and so this has virtual dimension $\mathrm{ind}^+ P$.

We have then for any instantons with any given flat limit ρ

Theorem 4.13 *For generic smooth metrics on X all instantons which are not flat are regular and for each bundle P the framed moduli space*

$$\tilde{M}_P \subset \tilde{\mathcal{B}}_P$$

is a smooth manifold of dimension $\mathrm{ind}^+ P + \dim H_\rho^0$ *with a smooth Γ_ρ action. The open set M_P^* in the instanton moduli space M_P is a smooth manifold of dimension* $\mathrm{ind}^+(P)$. *Moreover, if $b_X^+ \geq 1$ then for generic metrics on X there are no non-trivial reducible instanton solutions, so $M_P = M_P^*$.*

Of course we can go on to describe the structure around reducible solutions, just as in the compact case.

It is worth pointing out that we could have used these weighted spaces in the previous acyclic case. The moduli spaces will be unchanged so long as the weight α is less than the first eigenvalue δ. If we take $\alpha > \delta$, but not equal to an eigenvalue, we still get a good theory but the moduli spaces we describe will be different, they will be submanifolds of the true moduli space constructed above. In fact if we represent an instanton solution in temporal gauge over an end, so we have a family $\frac{dA_t}{dt} = L_\rho A_t + A_t \wedge A_t$, we can define an element $f_1(A)$ of the δ eigenspace of L_ρ by the limit of the L^2-orthogonal projections of $A_t/\|A_t\|$.

Then if α lies between δ and the next eigenvalue the moduli space $\mathcal{A}^\alpha/\mathcal{G}^\alpha$ can be identified with the zero set of the map f_1. Similarly, on the zero set of f_1 we get a map f_2 by projecting to the second eigenspace, when α lies between the second and third eigenvalues the moduli space will be the common zero set of f_1, f_2 and so on. In the case when there is a non-trivial isotropy group Γ_ρ the maps f_i are equivariant maps on $\tilde{M}$ which descend to sections of bundles over M. In the model case when $Y = S^3$ and there is a smooth conformal compactification $X \cup \{x_\infty\}$, the projection $f_1(A)$ can be identified with the *curvature* of A at the point x_∞, and the higher terms with the covariant derivatives of the curvature.

4.4 Gluing instantons

In this Section we consider the non-linear version of the additivity property of the index developed in Chapter 3. Thus we return to the family of Riemannian manifolds $X^{\sharp(T)}$, obtained by gluing a pair of ends $Y, \overline{Y}$ in a manifold X, depending on a parameter $T \geq 0$, and a bundle $P^\sharp$ over $X^{\sharp(T)}$ obtained by gluing the ends of P over X. We wish to compare instantons on X and $X^{\sharp(T)}$, when T is large. As before, the most important case will be when X is a disjoint union $X_1 \cup X_2$ and the two boundary components lie in the different pieces, and to simplify our notation we will restrict attention to this case, the changes required for the general case being quite trivial. One basic model of this set-up is the particular case when Y is a round 3-sphere and X_1, X_2 have conformal compactifications $\hat{X}_1, \hat{X}_2$. The manifolds $X^{\sharp(T)}$ are then diffeomorphic to the ordinary connected sum of X_1, X_2 and the parameter T can be regarded in another conformal model as the inverse of the exponential of the size of the 'neck'. Thus our discussion generalises the problem of describing instantons over a connected sum. Specialising still further, we can consider the case when $\hat{X}_2$ is the standard 4-sphere. Then the connected sum is conformally diffeomorphic to the original manifold $\hat{X}_1$, but under this diffeomorphism connections which are close to some fixed connection over $X^2 = S^4 \setminus \{\text{pt.}\}$ are taken to 'highly concentrated' connections over the fixed manifold $\hat{X}_1$ when T is large. So our discussion also generalises the problem of describing such 'concentrated connections' over a compact manifold.

A sizeable body of gluing results have been developed in these various contexts, going back to the seminal paper of Taubes [44], and using a variety of technical methods. Developing a full theory is a long and fairly arduous business – in part because of the weight of analytical detail and in part because of the diversity of hypotheses one may consider. We wish here to strike a balance, on the one hand giving all the essential details in the gluing results required in this book, but without on the other hand reproducing a lot of well-documented material. We will follow the general strategy used in [17], for the case of connected sums. We will show how the main constructions go over to the framework of tubes, but we will leave some of the detailed steps for the reader to check through. As before, we will divide our discussion into two parts: first, in the preliminary part of this Section, the easier acyclic case and then, in the Subsection following, the more general non-degenerate case where we encounter 'gluing parameters'. In the Appendix we will extend the discussion to certain classes of degenerate flat limits.

We fix attention then on the situation where the flat connection ρ over the common end Y of X_1, X_2 is acyclic (for simplicity we assume there are no other ends, although these would make no real difference to our discussion). We suppose A_1, A_2 are instantons on the bundles P_1, P_2 which are regular points in their moduli spaces, so the operators D_{A_i} have zero cokernel, and their kernels $H^1_i \equiv H^1_{A_i}$ have dimension

$$\dim H^1_i = \operatorname{ind}(P_i).$$

Our first goal is to show that in this situation there is an instanton on the bundle P over $X^{\sharp(T)}$, for large T, which is close to A_i over X_i (or, more precisely, over the large common open region in $X^{\sharp(T)}$ and X_i). We first construct a connection A_0 on the bundle P over $X^{\sharp(T)}$ which will be close to the desired solution. By Proposition 4.3 we can represent A_1 over the end $Y \times [0, \infty)$ of X_1 as $\rho + a_1$ where a_1 and all its derivatives decay exponentially along the cylinder. Choose a fixed cut-off function χ, with $\chi(t) = 1$ for $t < 0$ and $\chi(t) = 0$ for $t \geq 1$, and let A_1' be the flattened connection equal to A_1 outside the tube and to

$$\rho + (\chi(t - T - 1)a_1(t))$$

over the tube. Thus A_1' agrees with ρ over the region $Y \times [T, \infty)$ in the end and $F^+(A_i')$ is supported in the band $Y \times (T-1, T)$. Now it is clear that this ASD part of the curvature of A_1' decays exponentially with T. More precisely we have for any p, k

$$\|F^+(A_1')\|_{L^p_k} \leq C_{p,k} e^{-\epsilon T}.$$

We define a connection A_2' in a symmetrical fashion over X_2. Since the connections A_i' agree with ρ over the ends $[T, \infty) \times Y$, $[T, \infty) \times \overline{Y}$, and since $\Gamma_\rho = \pm 1$, they glue together in a canonical way to give a connection A_0 over $X^{\sharp(T)}$. The compatibility of the metrics on the manifolds clearly implies that

$$\|F^+(A_0)\|_{L^p_k(X^{\sharp(T)})} \leq C_{p,k} e^{-\epsilon T}. \tag{4.12}$$

Now, following Taubes' approach in [44], we seek a nearby solution $A_0 + a$ to the instanton equations over $X^{\sharp(T)}$. Thus we want to solve the equation

$$d^+_{A_0} a + (a \wedge a)^+ = -F^+(A_0), \tag{4.13}$$

with a small. The linearised version of this problem would be to solve $d^+_{A_0} a = -F^+(A_0)$, that is, to find a right inverse for the operator $d^+_{A_0}$, and this is essentially the problem we have already solved in Chapter 3. There is a slight additional complication that the connections A_i' which we are gluing together now depend on T. To get around this we observe that $D_{A_i'} - D_{A_i}$ is the operation of multiplication by a term which decays exponentially and whose L^2_1 to L^2 operator norm is $\mathrm{O}(e^{-\epsilon T})$. Thus when T is large a right inverse for D_{A_i} gives us right inverses Q_i for the operator $D_{A_i'}$ with operator norm independent of T:

$$\|Q_i \phi\|_{L^2_1} \leq C_i \|\phi\|_{L^2},$$

where C_i depends only on A_i.

We can now apply the construction of Chapter 3 using these Q_i to get, when T is large, an approximate right inverse P and true right inverse Q for $d^+_{A_0}$, with

$$P = Q(d^+_{A_0} Q)^{-1}.$$

We seek our solution to Equation 4.13 in the form $a = Q\phi$ so we have to solve the equation

$$\phi + (Q(\phi) \wedge Q(\phi))^+ = -F^+(A_0). \tag{4.14}$$

In turn we solve this equation using the contraction mapping principle in suitable function spaces. Abstractly, if we can find Banach spaces $(U, \| \; \|_U)$, $(V, \| \; \|_V)$ over $X^{\sharp(T)}$ such that

- Q defines a bounded map from U to V with operator norm bounded by C, independent of T,
- there is bounded multiplication

$$\|a \wedge \|_U \leq M \|a\|_V \|b\|_V,$$

with M independent of U

then the contraction mapping principle implies that Equation 4.14 has a solution ϕ when $\|F^+(A_0)\|_U$ is small enough (for example less than $(100MC^2)^{-1}$). Moreover the U norm of ϕ is of the same order as that of $F^+_{A_0}$, and ϕ is the unique small solution in this sense.

The simplest function spaces to use to obtain a solution are $U = L^2, V = L^4$. Plainly these satisfy the second condition above. For the first condition we observe that by the Sobolev embedding $L^2_1 \to L^4$ over the manifolds X_i the operators $Q_i : L^2(X_i) \to L^4(X_i)$ satisfy T-independent bounds. Now, going back to the calculation in the proof of Proposition 3.9 with $P(\phi) = \beta_1 Q_1(\phi_1) + \beta_2 Q_2 \phi_2$ the difference $D_{A_0} Q(\phi) - \phi$ is a sum of terms

$$\nabla \beta_i * Q(\phi_i)$$

and, just as for the L^2 norms in the proof of Proposition 3.9, the $L^4(X^{\sharp(T)})$ norm of this is bounded by

$$\|Q(\phi_i)\|_{L^4(X_i)} \|\nabla \beta_i\|_{L^\infty} \leq 2\epsilon(T) \|\phi\|_{L^2}.$$

It follows then that the L^2 to L^4 operator norm of Q is bounded independently of T, and we obtain L^2 solutions ϕ to Equation 4.14, giving L^2_1 solutions $a = Q\phi$ to Equation 4.13.

There are however many different possible choices of function space in which we can solve the equation. For example we can work in any spaces $U_T = L^p_k(X^{\sharp(T)})$ with $p \geq 2, k \geq 0$. Here, for definiteness, we will define the L^p_k norm by the sum of the L^p norms of the first k covariant derivatives. To establish the first property above we use the same argument as in the proof of Proposition 3.9, modified slightly if $k > 1$. We have inverses Q_i for $D_{A'_i}$ which satisfy a uniform L^p_k to L^p_{k+1} bound,

$$\|Q_i \phi\|_{L^p_{k+1}} \leq C_{p,k} \|\phi\|_{L^p_k}.$$

We define Q in the same fashion as before,

$$Q(\rho) = \phi_1 Q_1(\phi_1 \rho) + \phi_2 Q_2(\phi_2 \rho),$$

using the same cut-off functions ϕ_i. Then the $L^2_k(X^{\sharp(T)})$ norm of the error term $D_A Q\rho - \rho$ is estimated by the sum of the $L^p_k(X_i)$ norms of $\nabla\phi_i * Q(\phi_i\rho)$. The crucial point here of course is the fact that we can compute norms equally either in $X^{\sharp(T)}$ or in X_i, and this is an obvious consequence of the fact that the identification between the manifolds is an isometry.

Now observe, extending the discussion in the proof of Proposition 3.9, that the C^0 norm of any iterated derivative of ϕ_i tends to zero as T tends to infinity. This means that we have

$$\|\nabla\phi_i * Q_i\phi_i\rho\|_{L^p_k} \leq c_T \|Q_i(\phi_i\rho)\|_{L^p_k},$$

where c_T tends to zero as $T \to \infty$. Thus

$$\|\nabla\phi_i Q_i(\phi_i\rho)\|_{L^p_k} \leq c_T c_{p,k} \|\phi_i\rho\|.$$

Then by the same bound on the derivatives of ϕ_i we have

$$\|\phi_i\rho\|_{L^p_k} \leq a_T \|\rho\|_{L^p_k},$$

with $a_T \to 1$ as $T \to \infty$. Putting these inequalities together we see that the L^p_k operator norm of $D_A Q - 1$ tends to zero as T tends to infinity, so we can modify Q to obtain a right inverse P satisfying the uniform estimate above as an operator from L^p_k to L^p_{k+1}.

As for the uniform multiplication property, this follows from the uniformity of the Sobolev constants on $X^{\sharp(T)}$. We need to see that the Sobolev inequalities

$$\|f\|_{L^q(X^{\sharp(T)})} \leq C_{p,q,k} \|f\|_{L^p_k(X^{\sharp(T)})}$$

with $q = 4p/(4 - pk)$ and

$$\|f\|_{L^\infty(X^{\sharp(T)})} \leq C_{p,k} \|f\|_{L^p_k(X^{\sharp(T)})}$$

with $pk > 4$, hold with constants independent of T. This uniformity follows from the fact that the Sobolev inequalities hold on the non-compact manifolds X_i (Proposition 3.20). Indeed if we choose a suitable partition of unity $\beta_1 + \beta_2 = 1$, for example $\beta_i = \phi_i^2$, and apply the Sobolev inequalities over X_i to $\beta_i f$, we obtain the desired estimate over X.

We see then that we can solve Equation 4.14 without difficulty in any of these function spaces. This is precisely the advantage of Floer's approach, using tubes, over the approaches (in the case when $Y = S^3$) using other conformal models, in which the choice of function space is a relatively subtle matter.

Standard bootstrapping (at least starting with $k > 0$ or $p > 2$) shows that the solution A we have constructed is smooth, and this is then the instanton over $X^{\sharp(T)}$ for large T, made by 'gluing together' A_1 and A_2.

The construction applies equally well in a family. Let N_1, N_2 be precompact subsets of the moduli spaces M_{P_1}, M_{P_2} all of whose points are regular. For large T we obtain a map

$$\tau = \tau_T : N_1 \times N_2 \to M_P,$$

by applying this construction to $A_i \in N_i$. To complete the picture we need to address two problems:

(1) Show that, for $T \gg 0$, τ is a diffeomorphism to its image, and that the points in the image are regular points.
(2) Describe the image of τ, and show to what extent 'all' instantons on $X^{\sharp(T)}$ for large T are obtained in this way.

As we have stated above, the discussion of these follows that for the case of connected sums quite closely, so we will be fairly brief.

We begin with point (1). This naturally breaks into two parts, the first being to show that the derivative of τ is an isomorphism. This is tied up with the additivity of the index in Chapter 3. On the one hand, if $d\tau$ is an isomorphism and the image point is regular then

$$i(P) = \dim H^1_A = \dim H^1_{A_1} + \dim H^1_{A_2} = \mathrm{ind}(P_1) + \mathrm{ind}(P_2).$$

On the other hand, if we apply the index result Proposition 3.9, it suffices to show only that A is regular and that $d\tau$ is injective. The fact that A is regular, i.e. that d^+_A is surjective, follows easily from the existence of the operator Q satisfying uniform estimates. Recall that we have $A = A_0 + a$, and our estimates give $\|a\|_{L^4} \leq Ce^{-\epsilon T}$. On the other there is a right inverse Q for $d^+_{A_0}$, with bounded L^2 to L^4 operator norm. It follows that the L^2 operator norm of $d^+_A Q - 1$ tends to zero and so d^+_A is surjective. Next, to show that the derivative $d\tau$ is injective we can consider restriction to some compact sets $G_i \subset X_i$, which we can regard as being simultaneously subsets of X. Thus we can restrict connections on P and P_i to the same space $\mathcal{B}_{G_i}$. When T is large the composite of τ_T with the restriction maps from M_P is C^1-close to the restriction maps from N_i to $\mathcal{B}_{G_i}$ and it follows readily that τ_T is an injective immersion, for large T.

We now turn to problem (2). We fix function spaces with norms $\| \|_U$ (on 2-forms) and $\| \|_V$ (on 1-forms), satisfying the two conditions above. We would like to show first that there is a constant κ such that for large T any instanton $A^\sharp$ over $X^{\sharp(T)}$ which is of the form

$$A^\sharp = A_0 + b,$$

with $\|b\|_V \leq \kappa$, is gauge-equivalent to a solution $\tau_T(B_1, B_2)$ for points $[B_i]$ close to $[A_i]$ in M_{P_i}. (More precisely, $[B_i]$ will be of distance $O(\kappa)$ from $[A_i]$.) To do this we suppose first that the indices $\mathrm{ind}(P_i)$ are both zero, so the instantons A_i are isolated in their moduli spaces. What we obtain immediately from our construction is that if b lies in the image of Q, say $b = Q\phi$, then if κ is small and T is large the solution is the given one. This just reflects the fact that Equation 4.14 has a unique small solution. So we need to prove that $A^\sharp$ is *gauge-equivalent* to a connection of the form $A_0 + Q\phi$.

At this point we can forget the instanton condition. We would like to prove that, for suitable parameters, *any* $A + b$ with $\|b\|_V \leq \kappa$ is gauge-equivalent to a connection $A_0 + Q\phi$ with $\|\phi\|_U = O(\kappa)$. We can show this using the method of continuity, applied to the family $A + tb$ with $0 \leq t \leq 1$. This is fairly straightforward exercise; it is worked through in detail in [24] for the case of connected sums, although the case at hand is actually substantially simpler, due to the acyclic nature of A and consequent absence of a 'gluing parameter'. We have to show that the set of parameters t for which there exists a solution pair u_t, ϕ_t with

$$u_t(A + tb) = A + Q\phi_t,$$

with $\|\phi\|_V = O(\kappa)$, is both open and closed. To prove openness we use the implicit function theorem. Consider for example time $t = 0$. If $u_t = e^{\xi_t}$ the linearised equation is

$$tb = d_{A_0}\xi + Q\phi,$$

so the implicit function theorem gives a solution for small t provided that

$$d_A \oplus Q : \Omega^0_{X^{\sharp(T)}}(\mathfrak{g}_P) \oplus \Omega^+_{X^{\sharp(T)}}(\mathfrak{g}_P) \to \Omega^1_{X^{\sharp(T)}}(\mathfrak{g}_P)$$

is surjective. But it is easy to see that the *index* of this operator is minus $\mathrm{ind}(P)$ (the operator Q can be deformed to

$$(d^+_{A_0})^* \left(d^+_{A_0}(d^+_{A_0})^* + 1\right)^{-1}),$$

which is zero by the additive property and our assumption that $\mathrm{ind}(P_i) = 0$. On the other hand, using suitable function spaces, we can prove that, when T is large, the operator is *injective.* If $d_A\xi + Q\phi = 0$ then

$$\phi = -d^+_{A_0} d_{A_0}\xi = -[F^+_{A_0}, \xi].$$

Now suppose that d_{A_0} is injective on $\Omega^0(\mathfrak{g}_P)$ and moreover that we have a bounded multiplication, for $\zeta \in \Omega^0_{X\sharp(T)}(\mathfrak{g}_P), f \in \Omega^+_{X\sharp(T)}(\mathfrak{g}_P)$:

$$\|\zeta f\|_U \leq C\|d_{A_0}\zeta\|_V \ \|f\|_U, \tag{4.15}$$

with a constant C independent of T. Then combining this with the uniform bound on Q we get

$$\|d_A\xi\|_V = \|Q\phi\|_V \leq C\|\phi\|_U \leq C\|F^+_{A_0}\|_U\|d_A\xi\|_V.$$

Then, since $\|F^+_{A_0}\|_U \to 0$ as $T \to \infty$, this implies that, for large T, $d_{A_0}\xi$ and hence also ξ and ϕ are zero, as required.

To establish the uniform multiplication law Formula 4.15 we need a uniform Sobolev embedding,

$$\|\zeta\|_{L^\infty} \leq C\|d_{A_0}\zeta\|_V. \tag{4.16}$$

(Note that Formula 4.16 obviously implies Formula 4.15 if we take $V = L^p$.) This uniform inequality is crucial to the proof. It is also the vital ingredient required to show that if b and $Q\phi$ are small then u_t is close to the identity, so we always stay in a regime in which the non-linear problem is well approximated by its linearisation at $t = 0$.

Lemma 4.14 *If $pk \geq 4$ then there is a constant $C_{p,k}$, independent of T, such that*

$$\|\zeta\|_{L^\infty(X\sharp(T))} \leq C_{p,k}\|d_{A_0}\zeta\|_{L^p_k(X\sharp(T))}.$$

Again the proof follows rather trivially from the corresponding inequalities on the non-compact manifolds X_i.

Corollary 4.15 *If $pk \geq 4$ then there are constants $C_{p,k,j}$, for $0 < j < k$, such that*

$$\|\zeta f\|_{L^p_j(X\sharp(T))} \leq C_{p,k,j}\|\zeta\|_{L^p_k}\|f\|_{L^p_j}.$$

This follows from expanding out terms in $\nabla^r(\zeta f)$, using Hölder's inequality, and the previous results.

Armed with Lemma 4.14 and Corollary 4.15 it is a simple matter to carry through the continuity argument. The openness, with suitable choices of parameters, follows from Inequality 4.15, and its extension to connections near A_0. For the closedness we obtain *a priori* bounds on a solution. Here it is perhaps most convenient to choose the norms,

$$\|f\|_U = \|f\|_{L^p}, \quad \|b\|_V = \|b\|_{L^q} + \|d^+_{A_0} b\|_{L^p},$$

with $q = 4p/4 - p$, although this is by no means essential. Then for suitable κ and T, we get an L^p bound on $F^+(A_0 + b_t)$. This gives, by gauge invariance, an L^p bound on $F^+(A_0 + Q\phi_t)$ and this in turn gives an *a priori* bound on the L^p norm of ϕ.

We are now within sight of our goal in this section: a general 'gluing theorem' in the acyclic case. There are two more observations to make. First we should return to discuss the more general situation when $\mathrm{ind}(P_i)$ need not be zero. We choose local co-ordinates

$$\chi_i : U_i \subset M_{P_i} \to R^{n_i}$$

around given points in the moduli spaces M_{P_i} using functions which depend on the restriction of connections to compact sets $G_i \subset X_i$ – i.e. the χ_i extend to $\mathcal{B}_{G_i}$. Then we can regard the functions χ_i as being functions also on the moduli space M_P, and we can consider the 'cut down' moduli space

$$L = \chi_1^{-1}(y_1) \cap \chi_2^{-1}(y_2) \subset M_P$$

which has virtual dimension 0. We can work with this cut down space just as we did with the full instanton moduli space in the previous case and we can show that there is a unique point in L represented by connections close to a connection A_0 over $X^{\sharp(T)}$ formed by gluing together $A_1 = \chi_1^{-1}(y_1) \in N_1$ and $A_2 = \chi_2^{-1}(y_2) \in N_2$.

Finally we will firm up the notion of connections $A^\sharp$ over $X^{\sharp(T)}$ which are 'close' to connections A_i over X_i. One notion is that of connections which are close, in the norm $\| \;\|_V$, to the glued connection A_0. Another notion is that of connections which are close to the A_i over some compact subsets G_i in X_i. For instantons these are essentially equivalent. For given G_i and q as above let us define

$$d_{\mathrm{comp}}(A^\sharp; A_1, A_2) = \inf \|u(A^\sharp) - A_1\|_{L^q(G_1)} + \inf \|u(A^\sharp) - A_2\|_{L^q(G_2)}$$

where the infimum is taken over all gauge transformations u. Also, with say the norm $\| \;\|_V$ given above, we put

$$d(A^\sharp; A_1, A_2) = \inf \|u(A^\sharp) - A_0\|_V.$$

It is clear that d_{comp} is controlled by d. On the other hand we have

Proposition 4.16 *Let A_i be instantons over X_i as above, with acyclic limits. There are compact subsets $G_i \subset X_i$ and constants $\delta, C > 0$ such that for any instanton solution A' on the bundle P over $X^{\sharp(T)}$ which satisfies $d_i(A', A_i) \leq \delta$ we have*

$$d(A^\sharp; A_1, A_2) \leq C d_{\text{comp}}(A^\sharp; A_1, A_2).$$

To prove this we choose the sets G_i and δ so that any instanton over $X^{\sharp(T)}$ which satisfies the condition stated has very small energy over the complement $X^{\sharp(T)} \setminus (G_1 \cup G_2) \cong Y \times (-T, T) \subset X^{\sharp(T)}$. Then we can apply Proposition 4.4.

Putting the work in this Section together we obtain

Theorem 4.17 *Let N_i be compact sets of regular points in the moduli spaces M_{P_i}. For any sufficiently small $\kappa > 0$, and for sufficiently large T, there are neighbourhoods V_i of N_i and a smooth map*

$$\tau_T : V_1 \times V_2 \to M_{P, X^{\sharp(T)}},$$

such that, for suitable compact sets $G_i \subset X_i$,

- *τ_T is a diffeomorphism to its image, and the image consists of regular points,*
- *$d_{\text{comp}}(\tau_T(A_1, A_2); A_1, A_2) \leq \kappa$ for all A_i in U_i,*
- *any connection $A^\sharp$ in M_P with $d_{\text{comp}}(A^\sharp; A_1, A_2) \leq \kappa$ for some $A_i \in N_i$ lies in the image $\tau_T(V_1 \times V_2)$.*

4.4.1 Gluing in the reducible case

We will now modify the constructions of the earlier part of this Section to include the case when the limiting flat connection ρ is reducible, with non-trivial stabiliser Γ_ρ and Lie algebra H^0_ρ. Of course this includes the model example of a connected sum. (This more general gluing theorem is not needed for the basic Floer theory, so some readers may prefer to skip it.) We retain the assumption that the flat limit is *non-degenerate*.

The problem we need to face is that the analytical setting for constructing the instanton moduli spaces on the individual 4-manifolds X_1, X_2 now involves weighted norms, and we need to see how this affects the gluing discussion. We will explain two approaches to this. The first is to observe that in reality, as far as the construction of solutions goes, the difficulty is illusory. The problem of constructing solutions involves finding right inverses for the d^+ operators whereas, in this situation, the weights are only needed because of the fact that the covariant derivative of our model over the tube has a non-trivial kernel. Thus, in the same vein as our discussion in Subsection 3.3.3, the phenomena have to do with *different* terms in the deformation complex

$$\Omega^0_X \to \Omega^1_X \to \Omega^+_X.$$

In line with this, and related to the discussion in Subsection 3.3.3, one can proceed by putting weights separately on the terms in the deformation complex. Over a tubular end $Y \times (0,\infty)$ we write the 1-forms as

$$\Omega^1_X = \Gamma(\Lambda^0_Y \oplus \Lambda^1_Y).$$

Then we use L^p norms defined by exponential weights on the terms Ω^0_X and $\Gamma(\Lambda^0_Y) \subset \Omega^1_X$, but put no weights on the terms Ω^+_X and $\Gamma(\Lambda^1_Y) \subset \Omega^1_X$. A discussion like that in Subsection 3.3.3 shows that this leads to a Fredholm theory while the weights do not enter into the norm on the 2-forms, which is the crucial thing in the gluing construction.

Alternatively we can proceed as follows, exploiting the exponential decay of the curvature of instantons with the given flat, non-degenerate, limit. We form the manifold $X^{\sharp(T)}$ as before, and an approximate solution A_0, made by gluing together A_1' and A_2'. Notice that there is now a 'gluing parameter' involved in defining A_0, arising from the stabiliser Γ_ρ, but for the moment we fix this. Now we define a function W_T on $X^{\sharp(T)}$ with

$$W_T = e^{\alpha(T-|t|)}. \tag{4.17}$$

Here we are identifying the connecting region in $X^{\sharp(T)}$ with $Y \times (-T,T)$ in the obvious way. Over the manifolds-with-tubular-ends X_1, X_2 we have fixed weight functions W_1, W_2. So over the connecting region $W_1 = e^{\alpha(T+t)}$ and $W_2 = e^{\alpha(T-t)}$. We observe that W_T is dominated by both W_1 and W_2 in the regions of common definition. Now we take the norms $\| \ \|_U$, $\| \ \|_V$, on 2-forms and 1-forms respectively, to be the L^p and L^p_1 norms defined by the weight function W^T.

The first issue is to construct an approximate right inverse to $d^+_{A_0}$. The Fredholm theory for the operators D_{A_i}, or equally the operators $D_{A_i,\underline{\alpha}}$ of Subsection 3.3.3, gives (assuming initially as usual that the relevant groups $H^2_{A_i}$ vanish) inverses Q_i, bounded with respect to the W_i-weighted norms over X_i. We perform the 'splicing' construction a little differently from that in Chapter 3, setting

$$P(\rho) = \gamma_1 Q_1(\psi_1\rho) + \gamma_2 Q_2(\psi_2\rho).$$

Here, over the connecting region $Y \times (-T, T)$, the function ψ_1 is the characteristic function of $(-T, 0)$ extended in the obvious way – by 1 over the 'X_1 side' and 0 over the 'X_2 side'. The function γ_1 is chosen to be equal to 1 over the support of ψ_1 and with $|\nabla\gamma_1| = O(T^{-1})$. Symmetrically for ψ_2 and γ_2. Now we have $\gamma_1\psi_1 + \gamma_2\psi_2 = 1$ everywhere, so it follows that

$$d^+_{A_0} P(\rho) = \nabla\gamma_1 * Q_1(\psi_1\rho) + \nabla\gamma_2 * Q_2(\psi_2\rho).$$

The W_T-weighted norm of $\psi_1\rho$ is equal to the W_1-weighted norm (since the weight functions are equal on its support). Thus

$$\|Q_1(\psi_1\rho)\|_{L^{p,\underline{\alpha}}_1} \leq \text{const.}\, \|\rho\|_U.$$

On the other hand, the W_T-weighted norm of

$$\nabla\gamma_1 * Q_1(\psi_1\rho)$$

is *less than* the W_1-weighted norm (since the weight function is smaller), so we conclude that

$$\|\nabla\gamma_1 * Q_1(\psi_1\rho)\|_U \leq \text{const.}\, T^{-1}\|\rho\|_U,$$

and we can proceed as before. The bounded multiplication property

$$\|a \wedge b\|_U \leq M\|a\|_V\|b\|_V$$

follows from the same unweighted case using the arguments of Proposition 3.23 and Formula 4.10. The only other point is to see that the initial error term $F^+(A_0)$ is small in the U norm, when T is large. The problem here is that the weight function W is very *large* over the support of $F^+(A_0)$, in fact $O(e^{\alpha T})$. However, the exponential decay of curvature (Formula 4.12) tells us that the L^∞ norm of $F^+(A_0)$ is $O(e^{-\delta T})$ so

$$\|F^+(A_0)\|_{L^{p,\underline{\alpha}}} = O(e^{(\alpha-\delta)T}),$$

and this is small provided we fix $\alpha < \delta$. With these remarks in place, the construction of the glued solutions goes through as before. We now turn to the gluing parameter; there is a family of choices giving a family of gauge equivalence classes of connections A_0 parametrised by

$$Gl_{A_1,A_2} = \Gamma_\rho/(\Gamma_{A_1} \times \Gamma_{A_2}).$$

The global situation in moduli space is now rather more complicated because we may encounter reducible connections A_i and the isotropy groups Γ_{A_i} may change. Let $\tilde{M}_i$ be the based moduli spaces defined in Section 4.3. The group Γ_ρ acts on $\tilde{M}_i$ with quotient M_i and isotropy groups Γ_{A_i}. Let $\pi_i : \tilde{M}_i \to M_i$ be the quotient map, N_i be pre-compact sets of regular points in M_i and $\tilde{N}_i = \pi_i^{-1}(N_i)$. Our construction gives then smooth maps

$$\tilde{\tau}_T : \tilde{N}_1 \times N_2 \to M_{P1},$$

which are constant on Γ_ρ orbits, and so induce maps

$$\tau_T : E = \tilde{N}_1 \times_{\Gamma_\rho} \tilde{N}_2 \to M_{P1}.$$

The form of E will vary depending on the isotropy. If there are no reducible connections in the N_i then E is the total space of a $\Gamma_\rho \times \Gamma_\rho$ bundle over $N_1 \times N_2$, with with fibres $\Gamma_\rho / C(G)$. The arguments used before, modified as in [17][Chapter 7], go over to prove that, for large T, the map τ_T maps E isomorphically to a neighbourhood in M_{P1}.

4.5 Appendix A: further analytical results

In this Chapter we have developed the fundamentals of Yang–Mills theory over a 4-manifold with tubular ends in the case when the limiting flat connections are non-degenerate. This essentially suffices for setting up the Floer theory since, as we shall see in Chapter 5, one may always perturb the problem to achieve this non-degeneracy. However, there are good reasons for studying more general situations, in which the limiting flat connections are degenerate – certainly this often occurs for familiar 3-manifolds. This point of view has been developed extensively by Taubes [47] and Morgan, Mrowka and Rubermann [36]. For one thing, the ideas are useful in calculations in Floer theory; for another, they are essential in studying more geometrical aspects of the instanton moduli spaces – particularly in connection with algebraic geometry. Thus in this Appendix we will say a little about this more general situation.

4.5.1 Convergence in the general case

The first question we wish to address is the *limiting behaviour* of finite energy instantons over a general 4-manifold with tubular ends. Thus, fixing attention on one end, we may as well consider an ASD connection $\mathbf{A}$ over a half-tube $Y \times (0, \infty)$ with

$$\int_{Y \times (0,\infty)} |F(\mathbf{A})|^2 < \infty.$$

As usual, we let $[A_t] \in \mathcal{B}_Y$ denote the connection over Y obtained by restricting $\mathbf{A}$ to the slice $Y \times \{t\}$. Then, with no other assumptions on Y, we wish to prove

Theorem 4.18 *The connections $[A_t]$ converge (in the C^∞ topology on $\mathcal{B}_Y$) to a limiting flat connection $[A_\infty]$ over Y.*

First, the space $\mathcal{R}_Y \subset \mathcal{B}_Y$ of equivalence classes of flat connections is a finite union of disjoint path components $\mathcal{R}_Y^{(0)} \cup \cdots \cup \mathcal{R}_Y^{(N)}$ say. (Each is a real algebraic variety.) Our previous argument using the Uhlenbeck convergence theorem on bands shows that for any sequence $T_i \to \infty$ there is a subsequence $T_{i'}$ such that $[A_{T_{i'}}]$ converges to a flat limit. If we knew that we obtained the same limit for any sequence we would be done. The previous argument, for the case when the flat connections are isolated, extends immediately to show that there is some component, $\mathcal{R}_Y^{(0)}$ say, of the flat connections, such that

$$[A_t] \to \mathcal{R}_Y^{(0)} \text{ as } t \to \infty$$

(since the distance, in the L^2 metric, between the different components $\mathcal{R}_Y^{(\lambda)}$ is positive). The problem is to show that the path $[A_t]$ cannot 'spiral in' towards $\mathcal{R}_Y^{(0)}$, with many different limit points.

However if we know that

$$I = \int_0^\infty \|F(A_t)\|_{L^2(Y)} < \infty, \tag{4.18}$$

then this spiralling cannot occur and the path must have a unique limit point. For, by the instanton equation, the L^2 norm of the derivative of the path A_t in $\mathcal{B}_Y$ is $\|F(A_t)\|$. So, by the triangle inequality, for any sequence $t(i)$ the sum of the L^2 distances

$$\sum_i d([A_{t(i)}], [A_{t(i+1)}])$$

is bounded by I (the total length of the path). Since d is a metric this implies that there can be at most one limit point.

Our task then is to show the finiteness of the curvature integral Formula 4.18. (It is easy to show, using the elliptic estimates on bands, that this is equivalent to the condition that the curvature of $\mathbf{A}$ be in L^1 over the tube.) Observe first that the Chern–Simons function ϑ is constant on the path component $\mathcal{R}_Y^{(0)}$. We may suppose the restrictions A_t lie in some neighbourhood N of this component on which the Chern–Simons function can be lifted to a real-valued function and, without any loss, we may suppose that $\vartheta = 0$ on $\mathcal{R}_Y^{(0)}$.

We define a function f on $[0, \infty)$ by

$$f(t) = \|F(A_t)\|_{L^2(Y)}.$$

We know that f is in L^2 and we wish to show that f is in L^1. As before, we put

$$J(T) = \int_T^\infty |F(\mathbf{A})|^2.$$

So our Chern–Weil equality becomes

$$\frac{1}{8\pi^2} J(T) = \vartheta(A_t).$$

On the other hand we have

$$\frac{dJ}{dt} = -2\|F(A_t))\|_{L^2(Y)}^2.$$

The crux of our proof is an inequality relating the two invariants ϑ and $\|F\|$ of connections over Y.

Proposition 4.19 *There are a neighbourhood N of $\mathcal{R}_Y^0$, a constant λ with $1/2 \leq \lambda < 1$ and a $C > 0$ such that for any connection $[A] \in \mathcal{R}_Y^{(0)}$*

$$\vartheta(A)^\lambda \leq C\|F(A)\|_{L^2(Y)}.$$

We will now complete the proof of the Theorem, assuming this Proposition, before returning to the proof of Proposition 4.19. Clearly we may assume that the inequality applies to each A_t, so we have a differential inequality

$$\frac{dJ}{dt} \leq -C' J^{2\lambda}.$$

Of course, if $\lambda = 1/2$ we are in just the situation considered before, leading to exponential decay. Otherwise, we integrate this inequality to get

$$J(T) \leq C'' T^{-a}$$

for an exponent $a = \frac{1}{2\lambda - 1} > 1$. Now choose b with $a > b > 1$ and consider

$$\begin{aligned} S(T) &= 2\int_1^T f^2 t^b\, dt \\ &= \int_1^\infty \left(-\frac{dJ}{dt}\right) t^b\, dt, \\ &= \left(-t^b J(t)\right)_1^T + \int_1^T b t^{b-1} J(t)\, dt. \end{aligned}$$

We know that $J(t)$ is $\mathrm{O}(t^{-a})$ and we deduce then that $S(T)$ converges as $T \to \infty$, so

$$\int_1^\infty f^2 t^b\, dt < \infty.$$

Finally, write

$$f = (f\, t^{b/2})\, (t^{-b/2}),$$

and apply the Cauchy–Schwarz inequality to get

$$\int_1^\infty f\, dt \leq \left(\int_1^\infty f^2 t^b\, dt\right) \left(\int_1^\infty t^{-b}\, dt\right) < \infty$$

(since $b > 1$); thus we have shown that f is integrable, as required.

It remains then to establish the inequality of Proposition 4.19, relating the Chern–Simons invariant and the curvature. This follows from rather general considerations, using a method due (in another geometric problem) to L. Simon [42]. Consider first a finite-dimensional situation in which we have a function ϕ on an open set U in $\mathbf{R}^n$ whose critical set contains a component K, on which $f = 0$. Let $U' \subset\subset U$ be any (pre-compact) interior domain. We have then

Proposition 4.20 *If f is real analytic then there are constants A and μ with $1/2 < \mu < 1$, such that for all x in some neighbourhood of $U' \cap K$*

$$|f(x)|^\mu \leq A|\nabla f|.$$

This follows from fundamental structure theorems of Lojaszewicz for real analytic functions, see [34], [18]. For a very simple example take $n = 1$ and $f(x) = x^k$. Then $\nabla f = kx^{k-1}$ so we can take

$$\mu = \frac{k-1}{k}.$$

The inequality of Proposition 4.19 is clearly an analogue of this finite-dimensional result, since the curvature is identified with the gradient of the function ϑ, relative to the L^2 metric. On the other hand this infinite-dimensional case is easily reduced to the finite-dimensional result. First, since the critical set is compact, we can immediately reduce to proving the inequality on a small neighbourhood of any flat connection $[\rho] \in \mathcal{R}_Y^{(0)}$. Now we use the generalised Morse lemma of Appendix A to Chapter 2, so that we may assume the Chern–Simons function is represented on this neighbourhood as

$$\vartheta(x, y) = q(y) + h(x),$$

where h is a real analytic function on the finite-dimensional space H^1_ρ and q is a quadratic form on the orthogonal complement. The local diffeomorphism of $\mathcal{B}_Y$ which puts ϑ in this form is L^2-compatible so it suffices to prove that, for the gradient $\nabla'\vartheta$ defined relative to the *product* structure, we have

$$|\vartheta|^\mu \leq A|\nabla'\vartheta|,$$

for suitable constants. But $\nabla'\vartheta = (\nabla q, \nabla h)$ so

$$|\nabla'\vartheta|^2 = |\nabla q|^2 + \nabla h|^2.$$

The desired inequality therefore follows, since it holds for the non-degenerate quadratic function q and the finite-dimensional part h, by Proposition 4.20.

While we have obtained the general convergence result Theorem 4.18 by appealing to the abstract inequality Proposition 4.20, for specific 3-manifolds Y one can verify this directly, and also obtain more detailed results. An important case, which arises often in examples, occurs when the Chern–Simons functional is locally a 'Morse–Bott' function, i.e. when we can take the function h in the local model to be identically zero, and H^1_ρ represents the tangent space to the moduli space of flat connections at ρ. In this case we can take $\lambda = 1/2$ and our argument again give exponential convergence. We take this up in the next Subsection.

New phenomena, studied in detail in [47], [36], appear when we consider, for example, $SU(2)$ connections over the 3-torus $T^3 = S^1 \times S^1 \times S^1$. The variety of flat connections is a quotient of the Jacobian 3-torus J by the group of order 2, acting as ± 1. It is naturally decomposed into the smooth part J_{MB} representing the abelian flat connections with $\mathrm{ad}\,\rho$ non-trivial, around which the Chern–Simons functional is Morse–Bott, and the eight connections where $\mathrm{ad}\,\rho$ is trivial, which are also the fixed points of the action. Connections with limits in J_{MB} converge exponentially fast, and can be studied within the weighted space framework. The second case can be essentially reduced to considering connections with the trivial limit θ. One can write down a local model explicitly on the finite-dimensional space

$$H^1_\theta = \mathfrak{g} \otimes H^1(T^3; \mathbf{R}) = \mathfrak{g} \otimes \mathbf{R}^3 = \mathfrak{g} \times \mathfrak{g} \times \mathfrak{g}.$$

Here $\mathfrak{g}$ is the Lie algebra of the structure group, $SU(2)$. As in Appendix A to Chapter 2, the model is given by the function

$$h(\xi_1, \xi_2, \xi_3) = (\xi_1, [\xi_2, \xi_3]),$$

for $\xi_i \in \mathfrak{g}$. The special features here arise from the translations, $\mathbf{R}^3$, acting on the base manifold. The ordinary gradient flow of the function h on $\mathfrak{g} \times \mathfrak{g} \times \mathfrak{g}$ yields *Nahm's equations*

$$\frac{d\xi_i}{dt} = [\xi_j, \xi_k], \quad i, j, k \text{ cyclic.}$$

On the other hand one obtains these equations directly if one seeks T^3-invariant instantons on $T^3 \times \mathbf{R}$, in the form

$$\mathbf{A} = \xi_1\, dx_1 + \xi_2\, dx_2 + \xi_3\, dx_3.$$

It is not hard to verify that the inequality of Proposition 4.20 holds for this h with $\lambda = 2/3$. Then, by a simple extension of the proof of Proposition 4.19 we get

$$|F(\mathbf{A})| = \mathrm{O}(t^{-2+\delta})$$

for any $\delta > 0$ and instanton $\mathbf{A}$ over the tube. On the other hand we can write down explicit solutions which do not decay exponentially, and in fact with $|F(\mathbf{A})| = \mathrm{O}(t^{-2})$, through Nahm's equations. We just let $\eta_i \in (\mathfrak{g})$ satisfy the $SU(2)$ relation $\eta_i = [\eta_j, \eta_k]$, and set

$$\xi_i(t) = \frac{1}{t}\eta_i.$$

4.5.2 Gluing in the Morse–Bott case

It is obviously desirable to extend the gluing theory to more general situations, relaxing the non-degeneracy hypothesis, and here we will prove a simple result in this direction. We consider the Morse–Bott case, so we suppose that the variety of flat connections $\mathcal{R}_Y$ is a smooth r-manifold and the Chern–Simons functional is non-degenerate transverse to $\mathcal{R}_Y$. (The discussion is really local in $\mathcal{R}_Y$, so it suffices to assume this non-degeneracy locally.) We have seen then that any instanton on a bundle P_1 over a 4-manifold X_1 with end $Y \times [0, \infty)$ converges exponentially to a limit in $\mathcal{R}_Y$. The first item to discuss is the moduli theory of such instantons. For simplicity we suppose that all connections in $\mathcal{R}_Y$ are irreducible. (An important case when these assumptions hold is when Y is the product of a surface and a circle, and we work with non-trivial $SO(3)$ bundles.)

Fix a connection $\rho \in \mathcal{R}_Y$. We construct a moduli theory for instantons with limit ρ using weighted spaces with a small positive weight α. The entire discussion from the body of this Chapter goes through, and we get a moduli space $M_{X_1}(\rho)$ say, described as the zero set of a Fredholm section. The expected dimension of $M_{X_1}(\rho)$ is the index $\mathrm{ind}^-(P_1)$. Of course we can then define a larger moduli space M_{X_1}, as a set, as the union of the $M_{X_1}(\rho)$ as ρ varies over $\mathcal{R}_Y$. The essence of what is going on here is that the positive weight forces the variations in the connections that we are considering to decay at infinity, so the limit ρ is preserved. One might hope that if we used instead a small negative weight – allowing variations which grow at infinity – we would get a moduli theory in which the limit is allowed to vary, and so describe M_{X_1}, but unfortunately this does not work. The non-linear instanton equations do not behave well on the function spaces defined with negative weights. However, we can get around this by regarding ρ as a parameter, so we get a smoothly varying family of Fredholm moduli problems parametrised by $\rho \in \mathcal{R}_Y$. There is no difficulty in extending the usual discussion of local models to families and we arrive at the following picture. Let T denote the tangent space of $\mathcal{R}_Y$ at ρ and let $[A]$ be a point in $M_{X_1}(\rho)$. In the familiar way there are cohomology groups H^1_A, H^2_A defined by the deformation theory of A using the weighted spaces with small positive weight. There is a real analytic 'Kuranishi map'

$$\psi : T \times H^1_A \to H^2_A,$$

such that a neighbourhood of $[A]$ in M_{X_1} can be identified with a neighbourhood of 0 in

$$\psi^{-1}(0) \subset T \times H^1_A.$$

This means that we can endow M_{X_1} with the structure of a real analytic space. Moreover the expected dimension of M_{X_1} is

$$\dim T + \dim H^1_A - \dim H^2_A = \dim T + \mathrm{ind}^+(P_1) = \mathrm{ind}^-(P_1),$$

by Proposition 3.10. We say that a point A in M_{X_1} is a regular point if the derivative of the map ψ there is surjective. In this case M_{X_1} is locally a manifold of the expected dimension. By construction, M_{X_1} comes with a restriction map

$$r_{X_1} : M_{X_1} \to \mathcal{R}_Y.$$

We now move on to the gluing problem, so we suppose that we have another manifold X_2 with end $\overline{Y} \times (0, \infty)$ and form $X = X^{\sharp(T)}$ as before. The general idea is that a portion of the moduli space M_X is modelled on the *fibre product*

$$M_{X_1} \times_{\mathcal{R}_Y} M_{X_2} \equiv \{(A_1, A_2) \in M_{X_1} \times M_{X_2} : r_{X_1}(A_1) = r_{X_2}(A_2)\}. \tag{4.19}$$

That is, we can glue together instantons A_i over X_i provided that their limits are the same. Notice first that this is consistent with the linear gluing formula from Chapter 3 since (writing dim for the expected dimension)

$$\begin{aligned} \dim M_X &= \operatorname{ind}^+(P_1) + \operatorname{ind}^-(P_2) = \operatorname{ind}^-(P_1) + \operatorname{ind}^-(P_2) - \dim T \\ &= \dim M_{X_1} + \dim M_{X_2} - \dim \mathcal{R}_Y. \end{aligned}$$

We consider first the case when all the groups $H^2_{A_i}$ vanish, for all $A_1 \in M_{X_1}, A_2 \in M_{X_2}$. This means that the M_{X_i} are smooth manifolds and the restriction maps r_{X_i} are submersions. In particular the dimensions of the M_{X_i} are not less that the dimension of $\mathcal{R}_Y$. To make the gluing construction, given a pair of connections A_i over X_i with a common limit ρ, we choose a weight function W on the manifold $X^{\sharp(T)}$ just as in Equation 4.17. The gluing analysis in the weighted spaces goes through just as in Subsection 4.4.1, and we construct a family of solutions parametrised by the fibre product. We now move on to the more general case. The example to have in mind is when the M_{X_i} are smooth manifolds, but of lower dimension than $\mathcal{R}_Y$ and such that the maps r_{X_i} are *immersions*. In this case the expected dimension of each individual fibre $M_{X_i}(\rho)$ is negative and the fibres are generically empty. The fact that not all the fibres are empty means that we are forced to encounter non-trivial obstruction spaces $H^2_{A_i}$ and this prevents us from constructing an inverse operator Q satisfying uniform estimates. In fact to get a good gluing theorem in this situation we need an extra hypothesis.

Proposition 4.21 *Suppose* $(A_1, A_2) \in M_{X_1} \times M_{X_2}$ *with* $r_{X_1}(A_1) = r_{X_2}(A_2) = \rho$. *Suppose that*

- *the points* A_i *are regular points of the moduli spaces* M_{X_i} *(so these are locally smooth manifolds of the expected dimension),*
- *the map* $r_{X_1} \times r_{X_2} : M_{X_1} \times M_{X_2} \to \mathcal{R}_Y \times \mathcal{R}_Y$ *is transverse to the diagonal at* (A_1, A_2).

Then for large T *there is an open set in the moduli space* M_X *modelled on a neighbourhood of* (A_1, A_2) *in* $M_{X_1} \times_{\mathcal{R}_Y} M_{X_2}$.

For example, suppose that the M_{X_i} are compact and the r_{X_i} are injective immersions, so the moduli spaces M_{X_i} can be regarded as submanifolds of $\mathcal{R}_Y$. Then if these submanifolds intersect transversally the moduli space M_X can be identified (for large T) with the intersection $M_{X_1} \cap M_{X_2} \subset \mathcal{R}_Y$.

We can essentially reduce the proof of Proposition 4.21 to the case considered before. To simplify notation, let us suppose we are in the case when the map r_{X_2} is a local diffeomorphism whereas M_{X_1} consists of a single point A_1 with limit ρ. We can suppose then that for each σ near ρ there is a unique $A_2(\sigma) \in M_{X_2}$ with $r_{X_2}(A_2(\sigma)) = \sigma$ and the content of Proposition 4.21 is that for large T we can construct an instanton on $X^{\sharp(T)} = X$ by gluing A_1 and $A_2(\rho)$. Under the current hypotheses the dimension of $H^2_{A_1}$ is the same as that of $\mathcal{R}_Y$. We fix a set of compactly supported representatives $\chi_1, \ldots, \chi_r \in \Omega^+_{X_1}(\mathfrak{g}_P)$ for the obstruction space $H^2_{A_1}$. We want to consider an equation, for a connection A over X_1 and an r-vector $(\xi_1, \ldots, \xi_r) \in \mathbf{R}^r$:

$$F^+(A) + \sum_{p=1}^{r} \xi_p \chi_p = 0. \tag{4.20}$$

A little care is needed because this is not, as it stands, a gauge-invariant equation. One can construct similar gauge-invariant equations, by the devices used in [15] for example, but for the problem at hand we can get around the difficulty as follows. For each $\sigma \in \mathcal{R}_Y$ near to ρ we choose a connection $A_1 + a_\sigma$ on the fixed bundle P, with limit σ where a_σ varies smoothly with σ and $a_\rho = 0$. Of course the a_σ cannot decay along the tube, but we can require that $\|a_\sigma\|_{L^\infty} \to 0$ as $\sigma \to \rho$. We then consider connections in the form

$$A = A_1 + a_\sigma + b,$$

where b is in $L^{p,\alpha}_k$ and satisfies the gauge fixing condition $d^*_{A_1} b = 0$. Then we study Equation 4.20 for connections A of this form, so the equation is an equation for the triple $(\sigma, b, \underline{\xi})$. It follows then from the implicit function theorem that for each σ close to ρ there is a unique small solution $(b(\sigma), \underline{\xi}(\sigma))$ to Equation 4.20. Using this solution we can regard $\underline{i}$ as a function of σ and the ξ_p then give a set of local co-ordinates on $\mathcal{R}_Y$.

We now perform the gluing construction for each σ near to ρ. We glue the connection $A_1 + a_\sigma + b(\sigma)$ over X_1 to the connection $A_2(\sigma)$ over X_2. We obtain then a family of connections $A_0(\sigma)$ on a bundle P over $X = X^{\sharp(T)}$. Now we study the equation over X

$$F^+(A_0(\sigma) + c) + \sum_p \eta_p \chi_p = 0. \tag{4.21}$$

This makes sense because the bundle P is canonically identified with P_1 over the support of the χ_p. The construction of the inverse operator in the previous situation adapts easily to give a uniformly bounded right inverse (Q, q) to the operator

$$(c, \underline{\eta}) \mapsto d^+_{A_0} c + \sum \eta_p \chi_p.$$

Thus if we write $c = Q\phi$ we can apply the implicit function theorem in Banach spaces to the equation

$$F^+(A_0(\sigma) + Q\phi) + \sum_p \eta_p \chi_p = 0,$$

just as before, and we conclude that for each σ there is a unique small solution $(\phi(\sigma), \underline{\eta}(\sigma))$. Now $Q(\phi)$ is $O(e^{-\epsilon T})$ for some $\epsilon > 0$ and it follows that $\underline{\eta}(\sigma) - \underline{\xi}(\sigma)$ is also $O(e^{-\epsilon}T)$. Then when T is sufficiently small a standard degree argument, applied over a small ball in $\mathcal{R}_Y$ centred at ρ, shows that there is a solution σ to the equation $\underline{\eta}(\sigma) = 0$ close to ρ, and this yields the desired instanton over X. The rest of the proof of Proposition 4.21 follows familiar lines.

5

The Floer homology groups

This Chapter brings the first part of the book to its conclusion, with the construction of the Floer homology groups of a homology 3-sphere, using instantons over a 4-dimensional tube. Most of the technical work has been done in the previous two Chapters, but there are three further topics which we have kept for this Chapter. The first, which we take up in Section 5.1, is a discussion of *compactness* properties of instanton moduli spaces over manifolds with tubular ends. These are crucially important in Floer's theory, but the proofs are straightforward applications of the basic results summarised in Chapter 2. The next topic is the *orientation* of the moduli spaces or, better, of orientation line bundles formed from virtual index bundles. The key point here is a simple extension of the additive formula of Proposition 3.8. The other technical topic is a discussion of suitable *perturbations* of the instanton equation, which are constructed in Section 5.5. For purposes of exposition we give the main idea of Floer's theory at the earliest possible stage in this Chapter by working modulo 2 (which avoids orientations) and making a general position assumption (which avoids perturbations). These two extra topics are then fitted on to give the general definition of the Floer groups, using $SU(2)$ bundles over homology spheres. In the last Section 5.6 we discuss a straightforward extension of the theory to $SO(3)$ connections.

5.1 Compactness properties

As usual, we let X be a 4-manifold with tubular ends. We want to know what convergence properties hold for sequences of instantons over X with bounded energy. Recall that over compact subsets of X we have the general result from Chapter 2: any such sequence has a 'weakly convergent' subsequence. This weak convergence modifies the strong (C^∞) convergence that one would have for solutions of linear elliptic equations. The prototype example which shows that this modification is necessary is given by simply rescaling the standard instanton I_1 over R^4, i.e. considering a family of pull-backs by $\delta_\lambda : \mathbf{R}^4 \to \mathbf{R}^4$, $\delta_\lambda(x) = \lambda x$,

$$I_\lambda = \delta_\lambda^*(I_1).$$

The new feature we have to understand now derives from the fact that X is not compact – it has ends modelled on half-tubes. The most important case to have in mind is when X is itself a tube $Y \times \mathbf{R}$. On these tubes the *translations*

$$c_T : Y \times \mathbf{R} \to Y \times \mathbf{R};\ c_T(y,t) = (y, t+T)$$

play a role analogous to that of the dilations on $\mathbf{R}^4$ above. Indeed, in the model case when $Y = S^3$ the standard conformal equivalence $R^4 \setminus \{0\} \equiv S^3 \times \mathbf{R}$ takes the dilation $\delta_{\exp(T)}$ to the translation c_T. Under this conformal equivalence the weakly convergent family of instantons I_λ over $\mathbf{R}^4$, with curvature concentrating to a delta-function at 0, is taken over to a divergent family of instantons over the tube, with curvature sliding off to infinity. Just as the family of instantons I_λ was the basic model for the phenomenon of weak convergence this kind of sliding off provides the basic model for the new phenomena we encounter over tubes. There is a natural transversal for the action of the translations on the moduli spaces. We define a centred instanton on $Y \times \mathbf{R}$ to be one where

$$\int_{-\infty}^{\infty} t|F|^2 = 0,$$

that is, where the curvature density has centre of mass 0. Then we can define M'_P to be the moduli space of centred instantons and, except in the case when the instantons are flat (and hence constant in t), we have

$$M_P = \mathbf{R} \times M'_P. \tag{5.1}$$

We will sometimes refer to M'_P as the (translation-)*reduced* moduli space.

We begin with a result which will give convergence over cylindrical ends.

Lemma 5.1 *Suppose all flat connections over Y are non-degenerate. There are $\epsilon, \delta > 0$ such that if $[A_\alpha]$ is any sequence of instantons over*

$Y \times [0, \infty)$ with

$$\int_0^\infty |F(A_\alpha)|^2 \leq \epsilon,$$

then there are a subsequence α', a flat connection ρ and an instanton $\rho + a$ over the half-tube such that up to equivalence $A_{\alpha'} = A + a_{\alpha'}$, and for each p, k and $h > 0$,

$$\int_h^\infty |\nabla_A^{(k)}(a_{\alpha'} - a)|^p e^{p\delta t} \to 0$$

as $\alpha' \to \infty$.

This is a consequence of Proposition 4.3 on exponentially decaying gauges. Since there are only finitely many choices of ρ we may choose ϵ and a subsequence so that the $A'_{\alpha'}$ are represented as $\rho + a_{\alpha'}$, where all derivatives of the $a_{\alpha'}$ satisfy uniform exponential bounds. Then by Ascoli–Arzelà we may suppose that the $a_{\alpha'}$ converge pointwise to a limit a, and the dominated convergence theorem gives convergence in exponentially weighted L^p_k spaces.

Corollary 5.2 *Under the hypotheses above, a sequence of instantons $[A_\alpha]$ over $Y \times [0, \infty)$ converges in $L^p_{k,w}$ over a subdomain $Y \times [r, \infty)$, $r > 0$, to a limit $[A]$ if and only if $[A_\alpha]$ converges on compact subsets, and for some $h > 0$*

$$\int_h^\infty |F(A_\alpha)|^2 \to \int_h^\infty |F(A)|^2.$$

In particular, applying this to both ends, convergence in a moduli space M_P, defined by the L^p_k topologies, is equivalent to convergence over compact subsets:

Lemma 5.3 *A sequence $[A_\alpha]$ of connections on a given adapted bundle P over X converges to a limit $[A]$ in the same moduli space M_P if and only if the $[A_\alpha]$ converge in C^∞ on compact subsets of $Y \times \mathbf{R}$ to $[A]$.*

The point here is that the hypotheses imply that no curvature is 'lost' at the ends. In one direction, L^p_k convergence gives C^∞ convergence on compact subsets by local elliptic regularity for the instanton equations. In the other direction, by the Chern–Weil theory

$$\int_{Y \times \mathbf{R}} |F(A_\alpha)|^2 = \int_{Y \times \mathbf{R}} |F(A)|^2,$$

and we can apply Corollary 5.2.

We now make some useful definitions. By a *translation vector* $\underline{T}$ we mean a sequence of real numbers

$$T(1) < T(2) < \cdots < T(n).$$

Let $[A_\alpha]$ be a sequence of instantons on an adapted bundle P over the tube $Y \times \mathbf{R}$, and $\underline{T}_\alpha$ be a sequence of translation vectors, with $\underline{T}_\alpha(i) - \underline{T}_\alpha(i-1) \to \infty$ as $\alpha \to \infty$. We say $[A_\alpha]$ is $\underline{T}_\alpha$-convergent to limits $[A(1)], \dots, [A(n)]$, where the $A(i)$ are instantons over $Y \times \mathbf{R}$ (defined, in general, on different bundles), if for each i the translates

$$c^*_{T_\alpha(i)}(A_\alpha)$$

converge on compact subsets over the tube to $[A(i)]$. Similarly, we say that the connections are *weakly* $\underline{T}_\alpha$*-convergent* if the translates converge weakly on compact sets. In the latter case the limiting data will comprise connections $[A(i)]$ as above together with elements $Z(i)$ of the symmetric products of $Y \times \mathbf{R}$. Now let ρ^+ and ρ^- be the limiting flat connections of P at $\pm\infty$. We define a *chain* $\underline{P} = (P(1), \dots, P(n))$ of adapted bundles from ρ^- to ρ^+ to be a sequence of bundles with limiting flat connections $\rho^+(i), \rho^-(i)$ such that

$$\rho^- = \rho^-(1), \;\; \rho^+(1) = \rho^-(2), \dots, \rho^+(i) = \rho^-(i+1), \dots, \rho^+(n) = \rho^+.$$

We write $\kappa(\underline{P}) = \sum_{i=1}^n \kappa(P_i)$, so clearly $\kappa(\underline{P}) = \kappa(P) \bmod \mathbf{Z}$. We say that the sequence $[A_\alpha]$ is *chain-convergent* if there are a sequence $\underline{T}_\alpha$ of translation vectors, a chain of adapted bundles $P(1), \dots, P(n)$ as above and connections $A(i)$ on $P(i)$ such that the $[A_\alpha]$ are $\underline{T}_\alpha$-convergent to $A(1), \dots, A(n)$. Similarly we can define *weak chain convergence*, when the A_α are weakly $\underline{T}_\alpha$ convergent to $(A(1), Z(1)), \dots, (A(n), Z(n))$. Notice that, by Lemma 5.3, chain convergence with a chain of only one term ($n = 1$) is equivalent to a sequence of translates of the A_α converging in M_P.

It is very useful to see this chain convergence in the 3-dimensional picture of Chapter 2: viewing an instanton over the tube as a path in $\mathcal{B}_Y$. In this picture a chain-convergent sequence is a sequence of paths which dally for longer and longer periods close to the relevant flat connections.

We can now state our main compactness result in the form:

Theorem 5.4 *Any sequence* $[A_\alpha]$ *of connections on an adapted bundle* P *over* $Y \times \mathbf{R}$ *has a weak chain convergent subsequence. If the limit* $(\underline{P}, \underline{Z})$ *has* $\kappa(\underline{P}) = \kappa(P)$ *then the subsequence is chain-convergent.*

We choose ϵ as in Lemma 5.1. We may also suppose that any non-flat instanton over the tube has energy greater than ϵ, and that $\epsilon < 8\pi^2$, the energy of a 'single instanton'. We begin by constructing the first term P_1 in a limiting chain for a sequence A_α. For each α the energy of A_α is more than ϵ so there is a unique $T_\alpha = T_\alpha(1)$ such that

$$\int_{-\infty}^{T_\alpha} |F(A_\alpha|^2 = \epsilon.$$

Replacing the A_α by their translates we may without loss suppose that $T_\alpha = 0$ for each α. Now by our basic compactness result there is a subsequence of the A_α which is weakly convergent on compact subsets of the tube, and we may as well suppose this subsequence is the whole sequence. So $A_\alpha \to A(1)$ say, on compact sets, where $A(1)$ is an instanton on a bundle $P(1)$ over the tube.

We claim now that for any $h > 0$

$$\lim \int_{-\infty}^{-h} |F(A_\alpha)|^2 = \int_{-\infty}^{-h} |F(A(1))|^2. \tag{5.2}$$

This says that no energy is lost at the end in the limit. Note that we may not be able to take $h = 0$ if the A_α are weakly converging with a concentrated instanton at $t = 0$. On the other hand, since $\epsilon < 8\pi^2$ the connections are converging strongly over $(-\infty, -h)$. Suppose on the contrary that

$$e = \lim \int_{-\infty}^{h} \left(|F(A_\alpha)|^2 - |F(A(1))|^2\right)$$

is strictly positive. Note that $e \leq \epsilon$ by Theorem 5.4. Define $S_\alpha \leq -h$ by

$$\int_{-\infty}^{S_\alpha} |F(A_\alpha)|^2 = e.$$

Then it follows immediately from the strong convergence on compact sets that $S_\alpha \to -\infty$. Now let A'_α be the translates of the A_α by S_α. Taking a subsequence we can suppose that the A'_α converge, and the limit would then be an instanton with energy at most $e \leq \epsilon$, contradicting the definition of ϵ. Observe here that the A'_α cannot be only weakly convergent, since they have energy at most ϵ in the range $(0, -S_\alpha)$. This completes the proof of the energy equality, Equation 5.2.

Now we can apply Lemma 5.3 to see that the A_α are converging in L^p_k over the negative end, and in particular that the limiting flat connection $\rho^-(1)$ of $P(1)$ at $-\infty$ is equal to ρ^-.

Now it might happen that the energy of the ideal connection $A(1)$ is equal to the limit of the energy of the A_α. In that case we can apply Lemma 5.3 also to the positive end to see that the positive limit $\rho^+(1)$ of $P(1)$ is equal to ρ^+, and then we are done since the A_α are weakly chain-convergent to a chain with the single term $P(1)$. (It is also immediate that the convergence is strong if and only if $\kappa(P_1) = \kappa(P)$, i.e. if and only if P and $P(1)$ are isomorphic.) So suppose that

$$\kappa(P) = \kappa(A(1)) + \nu$$

with $\nu > 0$. Thus we 'lose' ν units of energy at the $+\infty$ end in the limit. Put

$$\epsilon' = \min(\nu, \epsilon)$$

(in fact, the argument below will show that $\nu \geq \epsilon$, so ϵ' is equal to ϵ). We define numbers $T_\alpha(2)$ by

$$\int_{-\infty}^{T_\alpha(2)} |F(A_\alpha)|^2 = \epsilon^-,$$

and consider the translates $A_\alpha(2) = c_{T_{\alpha(2)}}(A_\alpha)$. The same argument as before shows that $T_\alpha(2) \to +\infty$ and, after going to a subsequence, we can suppose the $A_\alpha(2)$ are weakly convergent on compact sets, to a non-trivial limit $A(2)$ on a bundle $P(2)$, with $\kappa(A(2)) \leq \nu$.

The task now is to see that the 'right hand' limit $\rho^+(1)$ of $P(1)$ is equal to the 'left hand' limit $\rho^-(2)$ of $P(2)$. For this we apply the decay result Proposition 4.4. There is an $\eta > 0$ such that for any $T > 1$ any instanton over $(-T, T)$ with energy less than η can be represented as $\rho + a$, for some flat connection ρ, such that for $|t| \leq T - 1$,

$$|\nabla^{(l)} a| \leq C_l e^{-\delta(t-T)}.$$

To use this we want to see that there is no energy 'lost' in the limit between $A(1)$ and $A(2)$. We can choose L such that for all large α the energy of A_α in the band $(L, S_\alpha - L)$ is less than η. In fact we may as well suppose that $L = 0$. Then we apply Proposition 4.4 to the translate $B_\alpha = c_{1/2S_\alpha}(A_\alpha)$. Transforming the bounds we see that there is some flat connection ρ over Y such that

$$\|A_\alpha|_\tau - \rho\| \leq Ce^{-\delta\tau},$$

while

$$\|A_\alpha(2)|_{-\tau} - \rho\| \leq Ce^{-\delta\tau},$$

and these immediately give that $\rho = \rho^+(1) = \rho^-(2)$.

Now, either the energies of $A(1)$ and $A(2)$ account for the total energy κ of the A_α or there is some remainder $\nu \geq \epsilon$. In the first case the argument above shows that the flat limit $\rho^+(2)$ is equal to ρ^+ and we are done. In the second case we use the same procedure to construct the third sequence of translates, and so on.

The process must terminate after a finite number of steps since each instanton $A(i)$ uses up more than ϵ energy, and this completes the proof of Theorem 5.4.

We now turn back to a general 4-manifold X with a tubular end $Y \times [0, \infty)$. (For simplicity we assume there is just one end.) It is easy to adapt the definitions and proofs above to this case. Let A_α be a sequence of instantons on an adapted bundle P over X, with flat limit ρ. We say the sequence is chain-convergent if there are a connection $A(0)$ on an adapted bundle $P(0)$ over X with limit $\rho(0)$, and a chain $\underline{P}$ from ρ^0 to ρ with instantons $A(i)$ on $P(i)$, such that

- the A_α converge to $A(0)$ over compact subsets of X,
- appropriate translates $c_{T_\alpha(i)}(A_\alpha)$ converge to the $A(i)$ on compact subsets of the tube,
- $\kappa(P) = \sum_0^n \kappa(P_i)$.

Here the only point to note is that in the second item each individual translate will only be defined on a half-tube, but since the union of these exhaust the whole tube as $\alpha \to \infty$ this does not affect the notion of convergence on compact subsets.

Similarly we define weak chain convergence, with weak convergence (to ideal instantons) replacing strong convergence over compact sets, and we have as before

Proposition 5.5 *Any sequence of instantons on P over X has a weak chain-convergent subsequence.*

There is a third situation we will want to consider, in which we have a sequence of connections A_α on a bundle P over a family of manifolds $X^{\sharp(T_\alpha)}$ as in Section 4.4, where the neck-length parameters $T_\alpha \to \infty$. We suppose $X^{\sharp(T_\alpha)}$ is formed by gluing a common end Y in manifolds X_1, X_2, which may have additional ends with cross-sections Y_λ. Let

ρ_λ be the flat limit of P over Y_λ. In this situation (if the curvature is bounded) the appropriate limiting data consists of

- instantons A^i on bundles P^i over X_i, with flat limits ρ^+, ρ^- over the ends $Y, \overline{Y}$, and σ_λ over the Y_λ,
- a chain of connections $\underline{A}$ over the tube $Y \times \mathbf{R}$, beginning at ρ^- and ending at ρ^+,
- chains of connections $\underline{A}_\lambda$ over tubes $Y_\lambda \times \mathbf{R}$ beginning at σ_λ and ending at ρ_λ.

We say that the connections $[A_\alpha]$ are chain-convergent to this limiting data if the connections converge to $[A^i]$ over compact subsets of X_i, and if appropriate translates of the restrictions of the A_α to the tube $Y \times (-T_\alpha, T_\alpha) \subset X^{\sharp(T_\alpha)}$ converge to the links $A(j)$ in the chain. Similarly for weak chain convergence. We again have a compactness theorem

Proposition 5.6 *In any sequence $T_\alpha \to \infty$ and sequence of instantons A_α on bundles $P \to X^{\sharp(T_\alpha)}$ there is a weak chain-convergent subsequence.*

Again, the proof is a minor variant of that of Theorem 5.4.

In our applications we will combine these basic compactness results with two other ingredients. The first is an immediate consequence of the additivity formula for the index. If we define the index of a chain $(\underline{P}, \underline{Z})$ to be the sum

$$\mathrm{ind}(\underline{P}, \underline{Z}) = \sum_j (\mathrm{ind}(P(j)) + 8|Z(j)|)$$

then we have the following Proposition, which merely spells out what has already been established in Chapter 3 in our present language.

Proposition 5.7 *If a sequence of connections A_α on a bundle P over $Y \times \mathbf{R}$ is weakly chain-convergent to a chain of ideal connections $\underline{A}, \underline{Z}$ on adapted bundles*

$$\mathrm{ind}(P) = \mathrm{ind}(\underline{P}, \underline{Z}) + H,$$

where H is the sum of the dimensions of the Lie algebras H^0_ρ of the vertices ρ of the chain.

Of course there are similar statements for connections over a manifold with tubular ends, or over a family of manifolds $X^{\sharp(T)}$.

The second ingredient is the gluing construction of Chapter 4. This provides, roughly speaking, a converse to the weak compactness results above. Again, there are a number of cases to consider, the basic case being that of a family of manifolds $X^{\sharp(T)}$. Given compact, regular, subsets N_i in moduli spaces M_{P_i} as in Section 4.4, we constructed a family $N_T = \operatorname{Im} \tau_T$ of instantons over $X^{\sharp(T)}$ for large T. If $T_\alpha \to \infty$ and we choose points $A_\alpha \in N_{T_\alpha}$ corresponding to fixed connections A^i in N_i (some choice may be involved if the limit is reducible) then A_α is chain-convergent to A^i in the sense above (with the trivial chain on the tube).

Now if X_1 and X_2 are both copies of a tube $Y \times \mathbf{R}$ then plainly $X^{\sharp(T)} = Y \times \mathbf{R}$ for each T. Thus our gluing results of Chapter 4 apply in this case to yield an 'addition' operation on instantons over tube. If we have a two-term chain of instantons $A(1), A(2)$, with $H^2_{A(i)} = 0$, then for large T we can construct an instanton on the tube which is close to the translate $c_{-T}(A(1)$ over the half-tube $Y \times (-\infty, 0)$ and to $c_T(A(2))$ over $Y \times (0, \infty)$. More precisely, given compact regular subsets N_i of the centred moduli spaces $M'_{P(i)}$ our construction gives us a family of connections N_T, an open subset in the centred moduli space $M'_{P(1)\sharp P(2)}$. In this language our results from Chapter 4 give

Proposition 5.8 *If $A(i)$ are instantons in N_i and A_α is a sequence of centred connections over $Y \times \mathbf{R}$ which is chain-convergent to $A(1), A(2)$ then for large enough α and suitable T, $[A_\alpha]$ lies in the subset N_T.*

Of course we can easily generalise this to longer chains, or more complicated kinds of convergence for example where there are limiting chains over the 'neck' in $X^{\sharp(T_\alpha)}$.

Let us complete this discussion of compactness properties by spelling out the particular conclusions we will need in the next Section. We consider a moduli space of instantons M_P on an adapted bundle P over a tube $Y \times \mathbf{R}$, and with non-trivial limits ρ^-, ρ^+ at either end. We suppose the metric on Y is such that all moduli spaces M_Q, for all bundles Q are regular. Then we have the **compactness principle**:

- If $\dim M_P < 9$ then any sequence in M_P has a chain-convergent subsequence. (That is, we do not lose 'concentrated instantons'.)
- If $\dim M_P < 5$ then any sequence in M_P has a subsequence converging to a limit $\underline{A}$ on a chain $\underline{P}$ of length $n \leq \dim M_P$ and with *non-trivial*

limits $\rho^-(j), \rho^+(j)$. (That is, the limit does not factor through the trivial connection.)

5.2 Floer's instanton homology groups

We can now explain the main construction in Floer's theory, in a simplified version in which we ignore orientation and make a general position hypothesis. We assume the following hypotheses.

- Y is a homology 3-sphere such that all irreducible $SU(2)$ connections over Y are non-degenerate.
- There is a metric on Y such that all moduli spaces of instantons on the Riemannian tube $Y \times \mathbf{R}$ are regular, i.e. all the groups H^2_A vanish for instantons A over the tube.

The first condition implies that the space $\mathcal{R}^*_Y$ of equivalence classes of irreducible flat connections is finite, and that we have a mod 8 degree

$$\delta_Y : \mathcal{R}^*_Y \to \mathbf{Z}/8,$$

as in Chapter 3. The *Floer chain group* C_* of Y (with $\mathbf{Z}/2$ co-efficients) is the $(\mathbf{Z}/2)$-vector space generated by $\mathcal{R}^*_Y$, i.e. for each $\rho \in \mathcal{R}^*_Y$ we have a basis element $\langle\rho\rangle \in C_*$. The map δ_Y defines a $(\mathbf{Z}/8)$-*grading*

$$C_* = \bigoplus_{i \in \mathbf{Z}/8} C_i,$$

with C_i generated by the $\langle\rho\rangle$ with $\delta_Y(\rho) = i$.

The (mod 2) Floer homology groups HF_i will be obtained from these chain groups by adding algebraic information extracted from the (reduced) moduli spaces of dimension 1 or 2. We define a linear differential

$$d : C_i \to C_{i-1},$$

in terms of the canonical bases

$$d(\langle\rho\rangle) = \sum n(\rho,\sigma)\langle\sigma\rangle$$

where the matrix elements $n(\rho,\sigma)$ are given as follows. If $\rho \in C_i, \sigma \in C_{i-1}$ and if P is an adapted bundle over the tube which is asymptotic to ρ at $-\infty$ and to σ at $+\infty$ then the index $\mathrm{ind}(P)$ is $\delta(\sigma) - \delta(\rho) = 1 \bmod 8$. There is thus a unique such bundle, up to equivalence, with $\mathrm{ind}(P) = 1$. By hypothesis the moduli space $M_P = M_P(g)$ is regular, with

$$\dim M_P = \mathrm{ind}(P) = 1.$$

However, the *translations* act freely on M_P, with quotient M'_P, so

$$\dim M'_P = 0,$$

and M'_P is a discrete set of points. According to the compactness principle of the previous Section, M'_P is compact (there can be at most one term in a limiting chain) so is a finite set. We put

$$n(\rho,\sigma) = \text{Number of points in } M'_P \mod 2.$$

We claim next that

$$d^2 : C_i \to C_{i-2}$$

is identically zero. Thus for each $\langle\rho\rangle \in C_i, \tau \in C_{i-2}$ we have to show that

$$\sum_{\langle\sigma\rangle \in C_{i-1}} n(\rho,\sigma)n(\sigma,\tau) = 0.$$

To see this we consider a bundle Q over the tube, asymptotic to ρ at $-\infty$ and to τ at $+\infty$ and with

$$\operatorname{ind}(Q) = 2.$$

Then we have a 1-dimensional reduced moduli space M'_Q of instantons on Q modulo translation. This moduli space M'_Q need not be compact, but we claim that the *ends* of M'_Q can be naturally identified with the pairs $(A(1), A(2))$ of connections on a two-term chain $P(1), P(2)$ from ρ to τ, so $\operatorname{ind}(P(1)) = \operatorname{ind}(P(2)) = 1$. This is an immediate consequence of the compactness principle together with Proposition 5.7. For each pair $A(1), A(2)$ we construct an open subset N_T in the moduli space M'_Q with a fixed parameter $T \gg 0$ using our gluing construction and Theorem 5.4 tells us that any divergent sequence in M'_Q eventually lies in one of these sets. It follows then that

$$\sum_{\langle\sigma\rangle} n(\rho,\sigma)n(\sigma,\tau)$$

is the number of ends of the 1-manifold M'_Q. Since a 1-manifold has an even number of ends this gives the required identity, working always modulo 2.

5.3 Independence of metric

From the definition given above, the Floer homology groups appear to depend on the metric chosen on Y (which defines the instantons on the tube). However, as we shall now show, the groups are in fact independent of this metric, up to canonical isomorphism. To show this we will develop the functorial property of the Floer groups with respect to cobordisms, described briefly in Chapter 1. This is one of the main topics in the book and we shall take the same ideas a good deal further in Chapters 6 and 7. Here we will just treat the class of 'h-cobordisms' which is quite sufficient for immediate applications. Thus we suppose that Y_0, Y_1 are two oriented homology 3-spheres and X is a 4-manifold with two-component boundary $\partial X = \overline{Y_0} \cup Y_1$, and such that the inclusions of each boundary component induce isomorphisms on homotopy groups. We assume that the 3-manifolds satisfy the hypotheses of the previous Section, for fixed Riemannian metrics g_i on Y_i. Thus we have Floer homology groups which we will temporarily denote by $HF_i(Y_0, g_0), HF_i(Y_1, g_1)$. Then we choose a generic metric G on the 4-manifold X_0, equal to the tubular metric formed from the g_i on collar neighbourhoods of each end.

We will show first that in this situation the Riemannian cobordism (X, G) defines linear maps on the Floer chain groups

$$\zeta_{X,G} : C_i(Y_0) \to C_i(Y_1).$$

We will see that these are functorial: if X' is a similar cobordism from $\overline{Y_1}$ to Y_2 and $(X^{\sharp(T)}, G^{\sharp(T)})$ is the Riemannian manifold formed by gluing across Y_1 with a long neck of length $\mathrm{O}(T)$, then for large enough T,

$$\zeta_{X',G'} \circ \zeta_{X,G} = \zeta_{X^{\sharp(T)},G^{\sharp(T)}} : C_i(Y_0) \to C_i(Y_2).$$

Also, the product cobordism induces the identity map. Moreover we will see that these maps $\zeta_{X,G}$ commute with the differentials defined above and hence induce maps on the Floer homology groups. Then we will show that these induced maps are independent of the metric G so we have

$$\zeta_X : HF_i(Y_0, g_0) \to HF_i(Y_1, g_1),$$

and these are also functorial with respect to the composition operation. The fact that the Floer groups of a 3-manifold Y are themselves independent of the metric is a formal consequence of these properties. For if g_0 and g_1 are two metrics on Y we can make a Riemannian cobordism with underlying manifold $X = Y \times [0, 1]$ and a metric G which agrees with g_0 at one end and g_1 at the other. This induces a map $\zeta_X : HF_i(Y, g_0) \to HF_i(Y, g_1)$, and similarly we get a map $\zeta_{X'} : HF_i(Y, g_1) \to HF_i(Y, g_0)$ using a cobordism X'. Then the composite $X \sharp X'$ can be deformed to the product metric on $(Y, g_0) \times [0, 1]$

– with the structure on the ends fixed throughout the deformation – so it follows from the properties above that $\zeta_{X'} \circ \zeta_X = \text{id}$, and similarly for $\zeta_X \circ \zeta_{X'}$.

Let us now work through these constructions in more detail. We extend X to a complete manifold with tubular ends, by adjoining tubes in the obvious way, and for simplicity we still call the resulting manifold X. For any pair of flat connections ρ over Y_0 and σ over Y_1 with

$$\delta_{Y_0}(\rho) = \delta_{Y_1}(\sigma)$$

there is an adapted bundle $P(\rho, \sigma)$ over X – unique up to isomorphism – with these flat limits at the ends and with

$$\text{ind}(P(\rho, \sigma)) = 0.$$

We define the maps $\zeta_{X,G}$ by counting instantons over X, with the given Riemannian metric G, in much the same way as we defined the boundary maps in the previous Section, making perturbations. For ρ and σ as above and for generic metrics G the moduli space $M_{P(\rho,\sigma)}$ is 0-dimensional, i.e. discrete, by construction. (Here we use the fact that the only flat bundles over X are those extending the flat bundles over the ends, which follows from the topological hypothesis.) The key fact we need is again the compactness of this moduli space.

Proposition 5.9 *If ρ, σ and $P(\rho, \sigma)$ are as above, the moduli space $M_{P(\rho,\sigma)}$ is compact.*

As usual, this is just a matter of analysing the possible limits using the compactness principles. For any sequence in $M_{P(\rho,\sigma)}$ we can pass to a subsequence which is weakly chain-convergent to limiting data consisting of

- an ideal instanton B, W on a bundle Q over X, having flat limits ρ', σ',
- a chain of ideal connections $\underline{A}_0, \underline{Z}_0$ over $Y_0 \times \mathbf{R}$, going from ρ to ρ',
- a similar chain $\underline{A}_1, \underline{Z}_1$ from σ' to σ over $Y_1 \times \mathbf{R}$.

The addition formula tells us that

$$\text{ind}(Q) + \text{ind}(\underline{A}_0) + \text{ind}(\underline{A}_1) + \dim H^0_{\rho'} + \dim H^0_{\sigma'} + 8(|\underline{W}| + |\underline{Z}_0| + |\underline{Z}_1|) \tag{5.3}$$

is equal to $\text{ind}(P(\rho, \sigma))$, and hence is zero. There are now two cases to consider. If the connection B is irreducible then its index $\text{ind}(B)$

is non-negative by general position. Since the index of any non-trivial chain is strictly positive we deduce that in fact $\rho = \rho'$, $\sigma = \sigma'$ and B is a limit point of the subsequence in $M_{P(\rho,\sigma)}$, as required.

Suppose on the other hand that B is a reducible instanton over X. Our assumptions on homology imply that B must be the trivial connection, which has index

$$-3(1 - b_1 + b_+) = -3.$$

On the other hand the isotropy algebras $H^0_{\rho'}, H^0_{\sigma'}$ have dimension 3 so, under our assumptions, we get

$$\mathrm{ind}(\underline{A}_0) + \mathrm{ind}(\underline{A}_1) \leq 3 - 3 - 3 < 0,$$

and this gives a contradiction.

We see then that $M_{P(\rho,\sigma)}$ is a finite set, and we put

$$N_{X,G}(\rho, \sigma) = \text{Number of points in } M_{P(\rho,\sigma)}.$$

Then we define

$$\zeta_{X,G}(\langle\rho\rangle) = \sum_{\langle\sigma\rangle} N_{X,G}(\rho, \sigma)\langle\sigma\rangle.$$

Now the fact that the product cobordism induces the identity map follows immediately from the action of translations, which means that non-flat instantons over the tube have index at least 1. The functorial property, for sums $X^{\sharp(T)}$ with $T \gg 0$, is an immediate consequence of the gluing result. Similarly the chain property

$$d_{Y_1} \circ \zeta_{X,G} + \zeta_{X,G} d_{Y_0} = 0 : C_i(Y_0) \to C_i(Y_1), \tag{5.4}$$

for large T, is obtained from an argument like that which we used to prove $d_Y^2 = 0$ above. We consider the matrix element, $D_{\rho,\tau}$ say, of the sum in Equation 5.4 corresponding to $\langle\rho\rangle \in C_i(Y_0), \langle\tau\rangle \in C_{i-1}(Y_1)$. By definition this is

$$D_{\rho,\tau} = \sum_{\sigma} n_{Y_0}(\rho, \sigma) N_X(\sigma, \tau) + \sum_{\sigma'} N_X(\rho, \sigma') n_{Y_1}(\sigma', \tau).$$

Now there is a bundle Q over X with flat limits ρ and τ and $\mathrm{ind}(Q) = 1$. We consider the 1-dimensional moduli space M_Q of instantons on Q. The same argument as we used before shows that any sequence in M_Q has a chain-convergent subsequence, with limit given by either

- an instanton on a bundle $P = P(\rho, \sigma)$, for some flat connection σ over Y_1, with $\mathrm{ind}(P) = 0$ and a one-term chain from σ to τ over $Y_1 \times \mathbf{R}$, or

- an instanton on a bundle $P' = P'(\sigma', \tau)$, for some flat connection σ' over Y_0, with $\text{ind}(P') = 0$, and a one-term chain from ρ to σ' over $Y_0 \times \mathbf{R}$.

Using the description of the ends given by the gluing theory we see that $D_{\rho,\tau}$ represents precisely the number of ends of M_Q, and hence is zero modulo 2.

It remains only to show that the map induced by $\zeta_{X,G}$ on the Floer homology groups is independent of the metric G. This is an instance of the general principle that 'counting instantons' or more generally 'counting solutions to an equation' should give an invariant independent of deformations. The argument refines that used in the basic case of compact 4-manifolds and for more general discussion of this case we refer to [17]. We will show that if $G(0), G(1)$ are two generic metrics on X, equal to the same product structures on the ends, then the maps $\zeta_0 = \zeta_{X,G(0)}, \zeta_1 = \zeta_{X,G(1)}$ are chain-homotopic. More precisely, given a generic one-parameter family of metrics $G(t)$ all fixed on the ends and interpolating between $G(0)$ and $G(1)$, we will define a chain homotopy

$$H : C_i(Y_0) \to C_{i+1}(Y_1),$$

such that

$$\zeta_1 - \zeta_0 = d_{Y_1} \circ H + H \circ d_{Y_0}. \tag{5.5}$$

Then, in the familiar way, this proves that ζ_0, ζ_1 induce the same map on homology. We define H by its matrix elements, in the usual way, so we fix attention on basis elements $\langle \rho \rangle \in C_i(Y_0), \langle \tau \rangle \in C_{i+1}(Y_1)$, and we consider the bundle R over X with limits ρ and τ and with

$$\text{ind}(R) = -1.$$

We know that for generic metrics on X the moduli space M_R is empty, since the index is negative. However in generic one-parameter families of metrics we do expect to encounter solutions. Thus, given our path $G(t)$ of metrics we let N be the moduli space of pairs (A, t) where A is an instanton on the bundle R relative to the Riemannian metric $G(t)$ on X. Then a familiar general position argument shows that, for generic one-parameter families $G(t)$, the space N is a discrete regular set, and an analysis of the possibilities as above shows that N is compact. (Here we need an obvious extension of our convergence results to families of metrics.) We can thus define the matrix element of H to be the number of points in this parametrised moduli space N.

Finally, then, we must see that this map satisfies the chain homotopy condition Equation 5.5. In the way which should now be quite familiar, this formula is distilled from the description of the end of an appropriate 1-dimensional moduli space. We fix attention on basis elements $\langle \rho \rangle \in C_i(Y_1), \langle \sigma \rangle \in C_i(Y_1)$ and the corresponding entry $l_{\rho,\sigma}$ in the matrix of the map $dH+Hd-(\zeta_1-\zeta_2)$. We need to show that we can obtain this number by counting the ends of an appropriate moduli space. We consider the moduli space L of pairs (A,t) as before, but now where A is an instanton on the bundle $P(\rho,\sigma)$, with index 0 and flat limits ρ, σ. Then for generic families $G(t)$ this moduli space L is a 1-dimensional manifold-with-boundary. The map $(A,t) \mapsto t$ gives a smooth map from L to the closed unit interval and the boundary of L is the pre-image of the end points, i.e. the union of the two moduli spaces $M_{P(\rho,\sigma)}(g_0), M_{P(\rho,\sigma)}(g_1)$ whose points define the matrix entries of ζ_1 and ζ_2.

We obtain the desired formula when we take account of the ends of L, in addition to the boundary points considered above. Suppose we have a sequence (A_α, t_α) in L, and $t_\alpha \to t$. Then, perhaps taking a subsequence, we may suppose that A_α either converges to a limiting $G(t)$-instanton or is chain-convergent to either

- a $G(t)$ instanton A on a bundle of index -1 over X, with limits ρ and σ' say, and a one-term chain over $Y_1 \times \mathbf{R}$ from σ' to σ, with index 1, or
- a one-term chain over $Y_0 \times \mathbf{R}$ from ρ to some ρ' with index 1, and a $G(t)$ instanton A on a bundle of index -1 over X, with limits ρ' and σ.

(Notice that in these cases we cannot have $t = 0, 1$ by our general position assumptions.) The final ingredient is to show that conversely if we are given a suitable $G(t)$ instanton A and the matching instanton I, say, over $Y_0 \times \mathbf{R}$ or $Y_1 \times \mathbf{R}$ then there exists a corresponding end of L. This is a straightforward extension of the gluing theory which we will now sketch briefly. (This is in the same vein as the gluing construction in the Appendix to Chapter 4.) If we fix t it may not be possible to construct a $G(t)$ instanton by gluing together A and I, since the cohomology group H^2_A is non-zero and this represents an obstruction to the gluing operation. In terms of our proof, we do not have the basic left inverse P_A to begin the construction. However, we do know that H^2_A is precisely 1-dimensional, and is generated by the time derivative

$$\frac{d}{ds}(F(A) + *_{G(s)}F(A))|_{s=t}.$$

This is just the condition that the parametrised moduli space be regular at (A,t). We represent the metrics $G(t+h)$ (or, rather, their conformal classes) relative to $G(t)$ by bundle maps $\mu_h : \Lambda^- \to \Lambda^+$ and set $M = \frac{d\mu_h}{dh}|_{h=0}$. Then the regularity condition can be expressed by saying that there is a linear operator $P : \Omega^+ \to \Omega^1$ and a linear functional π on Ω^+ such that

$$\rho = d_A^+(P\rho) + \pi(\rho)M(F^-(A)). \tag{5.6}$$

Now let us form flattened connections A', I' from A, I, as in Chapter 4, and a glued connection $\tilde{A}$, all depending on a parameter $T \gg 0$. We want to solve the equation for two variables (a,h):

$$F^+(\tilde{A}+a) + \mu_h(F^-(\tilde{A}+a)) = 0,$$

with a and h both small. The linearised version of this is to solve the equation

$$F^+(\tilde{A}) + d_{\tilde{A}}^+ a + hM(F^-(\tilde{A})) = 0.$$

In just the same way as before, one sees that one can solve the non-linear problem for large T if there are linear maps $\tilde{P}, \tilde{\pi}$ satisfying uniform estimates, such that

$$\rho = d_{\tilde{A}}^+ \tilde{P}(\rho) + \tilde{\pi}(\rho)M(F^-(\tilde{A})).$$

Now one shows first that small deformations $P_{A'}, \pi'$ of the solutions P_A, π of Equation 5.6 satisfy the corresponding equation

$$\rho = d_{A'}^+(P_{A'}(\rho)) + \pi'(\rho)M(F^-(A')).$$

The instanton I is a regular solution, so there is a right inverse P_I. Then we can glue together $(P_{A'}, \pi')$ and P_I to manufacture $(P_{\tilde{A}}, \tilde{\pi})$. The argument from Chapter 4 goes through without any essential change, since under our bundle identifications $M(F^-(\tilde{A})) = M(F^-(A'))$. (The map M is supported in the interior of X.)

In this way we see that there is exactly one end of L for each pair (A,I) of limiting data of the two types above. But, summing the number of these ends over all choices of σ', ρ', we see that (modulo 2) these two terms give exactly the matrix entries of $d_{Y_1}H$ and Hd_{Y_0} respectively. So, once again, the chain homotopy Equation 5.5 follows from the fact that

$$\text{Number of boundary points of } L + \text{Number of ends of } L = 0 \mod 2.$$

5.4 Orientations

We will now extend the theory of the previous Section to define integral Floer homology groups, or more generally Floer homology groups with co-efficients in any abelian group. For this we need to introduce orientations, and take due account of *signs*. We continue with our assumption that all flat limits we encounter are non-degenerate.

In Chapter 3 we have defined Fredholm operators associated to an adapted bundle P over a 4-manifold X with tubular ends. Specifically, here, we take the operators on unweighted spaces in the case of acyclic (i.e. irreducible) limits and those with small positive weights in the case of the reducible (i.e. trivial) limit. It follows then, just as in the compact case, that there is a determinant bundle $\lambda_P \to \mathcal{B}_P$ over the space of compatible connections. This is a real line bundle with the property that over a regular moduli space $M_P \subset \mathcal{B}_P$, the restriction of λ_P is isomorphic to the orientation bundle $\Lambda^{\max} TM_P$. Thus if λ_P is trivial over the space $\mathcal{B}_P$ of all connections the moduli space M_P is *a fortiori* orientable, and a choice of trivialisation of λ_P fixes an orientation of the moduli space. As a matter of language, in this Section, when we talk about trivialisations of line bundles and canonical isomorphisms between line bundles we are really referring to the orientation class, so our isomorphisms are canonical up to multiplication by positive scalars.

The fact that the line bundles λ_P are trivial follows from straightforward topological properties of the space $\mathcal{B}_P$. The whole discussion is, as usual, a variant of that in the compact case. Let us first digress to discuss briefly the homotopy type of this space. It is convenient to introduce also the space $\tilde{\mathcal{B}}_P$ of framed connections, fixing a frame over a base point in X. Suppose first that the flat limit of P over each end of X is *trivial.* Then we can regard P as being a bundle over the compactified space $\hat{X}$, as in Section 3.2. An argument just like that in the case of a compact base space X (see [17]) shows then that there is a natural homotopy equivalence

$$\tilde{\mathcal{B}}_P \simeq \mathrm{Maps}^*_P(\hat{X}, BG),$$

where BG is the classifying space of the structure group G under consideration (here $SU(2)$), and Maps^*_P denotes the homotopy class of based maps corresponding to the given bundle P.

On the other hand we can see that the homotopy type of $\tilde{\mathcal{B}}_P$ is independent of the given G bundle P by using the familiar addition operation. First, using a temporal trivialisation, it is easy to see that

$\tilde{\mathcal{B}}_P$ has as a deformation retract the subspace $\tilde{\mathcal{B}}_P^f$ of equivalence classes of framed connections which are flat over fixed tubular ends in X. The framing at the base point gives us then a definite isomorphism over the end with the standard flat model. Hence if X_1 and X_2 are two such manifolds with matching ends $Y, \overline{Y}$, and P_1, P_2 are bundles with matching limits over these ends, we get a gluing map

$$\sharp : \tilde{\mathcal{B}}_{P_1}^f \times \tilde{\mathcal{B}}_{P_1}^f \to \tilde{\mathcal{B}}_{P_1 \sharp P_2}^f. \tag{5.7}$$

Suppose in particular that $Y \times [0, \infty)$ is an end of X and P has flat limit ρ over Y. Fix another connection B on a bundle over the tube $Y \times \mathbf{R}$ with flat limits ρ at $-\infty$, σ at ∞. Then taking the connected sum with B gives a map

$$\sharp_B : \tilde{\mathcal{B}}_P^f \to \tilde{\mathcal{B}}_Q^f,$$

say, where Q is a bundle over X with flat limit σ. If B' is a connection on the 'reflected' bundle over the tube (switching $\pm\infty$) the composite map $\sharp_{B'} \sharp_B$ is homotopic to the identity on $\tilde{\mathcal{B}}_P^f$. To see this we just fix a path of connections over the tube beginning with $B \sharp B'$ and ending with the flat connection ρ, then glue this path to the connections over X. Thus we see that $\sharp_B$ induces a homotopy equivalence between $\tilde{\mathcal{B}}_P$ and $\tilde{\mathcal{B}}_Q$. (And at the level of homotopy this is independent of the connection B.)

This argument gives

Proposition 5.10 *For any adapted bundle P over X there is a homotopy equivalence*

$$\tilde{\mathcal{B}}_P \simeq \mathrm{Maps}_0^*(\hat{X}, BG),$$

where Maps_0^* *denotes the component of the trivial map.*

We now bring in the determinant line bundles. The essential point is the existence of an 'excision formula' generalising the addition property Proposition 3.9 for the numerical index. Let A_1 and A_2 be connections on bundles P_1, P_2 over manifolds X_1, X_2 as above, and let $A^\sharp = A_1 \sharp A_2$ be the connected sum connection, constructed with some large gluing parameter T.

Proposition 5.11 *There is a natural isomorphism*

$$\lambda_{P\sharp}(A^\sharp) = \lambda_{P_1}(A_1) \otimes \lambda_{P_2}(A_2)$$

This follows from our proof of the addition formula Proposition 3.9. We assume for simplicity that the connection ρ is irreducible: in fact this is the only case we need below, although the result is also true for the (trivial) reducible limit. In the simplest situation, where the operators D_{A_1}, D_{A_2} have trivial cokernels, we have constructed in Chapter 3 an isomorphism from $\ker D_{A_1} \oplus \ker D_{A_2}$ to $\ker D_{A\sharp}$. This isomorphism induces the desired isomorphism of determinant lines. The specific isomorphism of the kernels depends on various choices but it is easy to see from the construction that these are all homotopic and so induce the same isomorphism of determinant lines (up to positive scalars, as usual). In the more complicated case when the operators have non-trivial cokernels we get the required isomorphism from an exact sequence, just as in [17][Chapter 7].

We return now to the problem of trivialising these line bundles. The lift of λ_P to $\tilde{\mathcal{B}}_P$ is naturally an $\mathrm{ad}\, G$-equivariant line bundle $\tilde{\lambda}_P$ and so, since $\mathrm{ad}\, G$ is connected, it suffices to show that $\tilde{\lambda}_P$ is trivial. The argument for this, involving embedding the gauge group $SU(2)$ in $SU(n)$, goes exactly as in the case of compact 4-manifolds treated in [13]. Again, just as in the compact case, one can fix an isomorphism between these lines λ_P for adapted bundles with the same flat limits. If P, P' are two such bundles then P can be obtained from P' by adding instantons in small neighbourhoods of points in X, then one uses an excision argument again. Thus one gets in sum a line $\lambda_{X,\rho}$ associated to X and the flat limits. Moreover, if all the flat limits are trivial then the line λ_ρ can be identified with a line λ_X obtained from the real cohomology of X:

$$\lambda_X = \lambda_\theta \equiv \Lambda^{\max}(H^0(X) \oplus H^1(X) \oplus H^+(X)).$$

Now we claim that the line $\lambda_{X,\rho} \otimes \lambda_X^*$ is independent of the manifold X with boundary Y. This follows a familiar theme in this book – a pattern exemplified already by the definition of the Chern–Simons function and of the grading $\delta_Y(\rho)$. For suppose that X_1, X_2 are two choices and choose some 4-manifold W with (oriented) boundary $\overline{Y}$. Let $Z_1 = X_1 \sharp_Y W$ and $Z_2 = X_2 \sharp_Y W$. We choose connections A_i over X_i and B over W with flat limit ρ. Then we have two glued connections, $A_1 \sharp B, A_2 \sharp B$ say, over Z_1 and Z_2 respectively. Now our excision isomorphism of Proposition 5.11 gives isomorphisms

$$\lambda_{Z_i} \cong \lambda(A_i \sharp B)) \cong \lambda_{P_i}(A_i) \otimes \lambda(B) \cong \lambda_{X_i,\rho} \otimes \lambda_{W,\rho}, \tag{5.8}$$

while

$$\lambda_{Z_i} \cong \lambda_{X_i} \otimes \lambda_W. \tag{5.9}$$

The second isomorphism just follows from the standard Mayer–Vietoris isomorphisms

$$H^1(X) = H^1(X_1) \oplus H^1(X_2), \quad H^+(X) = H^+(X_1) \oplus H^+(X_2).$$

Putting together Formulae 5.8 and 5.9, we see that $\lambda_{X_i,\rho} \otimes \lambda^*_{X_i}$ is the same for either manifold X_1, X_2 – each is canonically isomorphic to $\lambda_{W,\rho} \otimes \lambda^*_W$. Thus we get a line

$$\lambda_\rho = \lambda_{X,\rho} \otimes \lambda_X$$

associated solely to the flat connections ρ. To sum up then, there is a line λ_ρ associated to each irreducible flat connection ρ such that trivialisations of λ_ρ and of the line λ_X yield a definite orientation of the determinant line associated to any bundle over a 4-manifold X, with homology orientation, and with boundary Y.

With all this background in place we can proceed to define the integral Floer chain groups. For a homology 3-sphere Y, satisfying the hypotheses of the previous Section, we fix a trivialisation of each line λ_ρ and decree that the symbol $\langle \rho \rangle$ is associated to ρ and this trivialisation. Then we define the integral chains $C_*(Y)$ to be the free abelian group generated by these symbols $\langle \rho \rangle$, with grading as before. In this group we can regard $-\langle \rho \rangle$ as being attached to ρ with the opposite trivialisation of the line. Another way to say this is, if we use real co-efficients, that

$$C_*(Y) = \bigoplus \lambda_\rho,$$

where the sum runs over the flat connections.

We next want to define the differential $d : C_i \to C_{i-1}$, by a sum

$$d\langle \rho \rangle = \sum n(\rho, \sigma) \langle \sigma \rangle,$$

as in Section 5.3, except now we need each matrix entry $n(\rho, \sigma)$ to be an integer. This integer is obtained by counting the points in the 0-dimensional moduli space $M'_{P(\rho,\sigma)}$ with suitable signs ± 1. Let A be an instanton on $P(\rho, \sigma)$. By our regularity hypothesis the cokernel of D_A is zero, and the kernel is generated by the action of the translations, so we can fix a canonical trivialisation of $\lambda_{P(\rho,\sigma)}(A)$. Now if we choose X as above we can glue A (after flattening the ends) to a connection on a bundle $P(\rho)$ with flat limit ρ to get a connected sum bundle $P(\sigma)$

with flat limit σ. Our addition formula of Proposition 5.11 gives an isomorphism

$$\lambda_{P(\sigma)} \cong \lambda_{P(\rho)} \otimes \lambda_{P(\rho,\sigma)}(A),$$

and then the trivialisation of the last term in this equation induces an isomorphism between $\lambda_{P(\sigma)}$ and $\lambda_{P(\rho)}$, and hence between λ_ρ and λ_σ. Finally then we get our sign $s(A) \in \{\pm 1\}$, for fixed symbols $\langle\rho\rangle, \langle\sigma\rangle$, by comparing the chosen trivialisations of $\lambda_\rho, \lambda_\sigma$ under this isomorphism. Then we define

$$n(\rho, \sigma) = \sum_{A \in M(\rho,\sigma)} s(A).$$

One now has to check that, with this definition of d_Y, we have $d_Y^2 = 0$ so we can can define Floer groups

$$HF_i(Y, \mathbf{Z}) = \frac{\ker d_Y : C_i \to C_{i-1}}{\operatorname{Im} d_Y : C_{i+1} \to C_i}.$$

Likewise one has to check that a Riemannian 4-manifold X with boundary $\partial X = Y_1 \cup \overline{Y}_0$ as in Section 5.3 induces a chain map ζ_X which, up to chain homotopy, is independent of the metric. These are all straightforward exercises, using oriented moduli spaces and oriented boundaries, which we leave for the reader. The basic point to note is that the gluing construction we have used to describe the ends of the relevant moduli spaces is compatible with the addition isomorphisms we have fixed in Proposition 5.11 for the indices. Notice that while there are a number of points where we have to fix conventions to define the signs $s(A)$ above these are not really important since a change in convention will just change the sign of d_Y and this does not affect the homology.

5.5 Deforming the equations

Up till now we have assumed that geometric data underlying the Floer theory is in 'general position' in that, first, the flat irreducible connections are acyclic and, second, the moduli spaces of instantons on the cylinder are regular (and so in particular are of the proper dimension). This Section is devoted to the extension of the theory in this case when these conditions do not hold. The approach is to recover the earlier picture by making a suitable small deformation of the problem: in spirit and technique this follows closely the similar discussion for 4-manifold invariants; cf. [13]. As usual we concentrate here on the case of $SU(2)$ connections but the arguments can be adapted, with minor changes, to other structure groups and to this end we formulate some of our results more generally.

We begin by describing the perturbations that we will allow. Let γ be an embedded loop in our homology 3-sphere Y, and extend the embedding to a solid torus $S^1 \times D^2 \to Y$. Thus for each point z in the disc we have a parallel copy γ_z of γ. If A is an $SU(2)$ connection on a bundle over Y we let $\tau_z(A)$ be the trace of the holonomy of the connection around γ_z. Now choose a smooth, positive, 2-form μ of compact support on the disc and with integral 1. We define

$$\sigma(A) = \int_{D^2} \tau_z(A)\mu(z). \tag{5.10}$$

This is plainly a gauge-invariant quantity: more precisely σ defines a smooth $\mathcal{G}$-invariant function, taking values in the interval $[-2, 2]$, on the space $\mathcal{A}$ of connections. More generally, if $\sigma_1, \dots, \sigma_N$ are functions formed in this way and $\eta : \mathbf{R}^N \to \mathbf{R}$ is smooth, we may consider a $\mathcal{G}$-invariant function

$$\underline{\eta}(A) = \eta(\sigma_1(A), \dots, \sigma_N(A)). \tag{5.11}$$

We call such functions $\underline{\eta}$ the *admissible* functions on $\mathcal{A}$. For any such function we may perturb the Chern–Simons function Φ to $\Phi + \underline{\eta}$.

Strictly we should now redo the entire differential-geometric and analytical theory for this perturbed situation, but in reality the previous discussion goes through with only small changes. We begin by considering the derivative of the function σ with respect to the connection A. This requires some notation. Let p be a point on a loop γ_z in Y. The holonomy around γ_z can be viewed as an element of the fibre of the adjoint bundle over p, a copy of $SU(2)$. Let $\pi : SU(2) \to \mathfrak{su}(2)$ be the projection given by

$$\pi(u) = u - \tfrac{1}{2}\mathrm{Tr}(u)1. \tag{5.12}$$

This is equivariant with respect to the adjoint actions of $SU(2)$ on itself and its Lie algebra, so induces a map from the adjoint bundle of groups to the adjoint bundle of Lie algebras $\mathfrak{g}_P$. Let $T_A(p)$ be the element of the latter bundle defined by applying this map to the holonomy. In this way we get a section T_A of the bundle g_P over the solid torus in Y, which has the property that its covariant derivative along the family of loops is zero. Now pull the 2-form μ back from the disc to the solid torus by the obvious projection. For simplicity we also denote this pull-back by μ. Plainly this can be viewed as a smooth 2-form on Y, supported in the solid torus.

Lemma 5.12 *The derivative of $\sigma(A)$ with respect to the connection A is the map*

$$a \mapsto -\int_Y \mathrm{Tr}(a \wedge (T_A \mu)).$$

Here the product $T_A\mu$ makes sense as a smooth bundle-valued 2-form over Y, even though T_A is only defined in the solid torus, because μ vanishes outside the solid torus. We will leave the proof of Lemma 5.12, which is largely a matter of notation, as an exercise.

We now turn to the critical points of the perturbed function $\Phi + \underline{\eta}$. Using the Lemma and the chain rule, these are solutions of the equation

$$F_A = \sum_{i=1}^{N} \partial_i \eta \; T_{A,i} \; \mu_i. \tag{5.13}$$

Here $\partial_i \eta$ denotes the partial derivative of $\eta : \mathbf{R}^N \to \mathbf{R}$ in the ith variable, evaluated at the point $(\tau_1(A), \ldots, \tau_N(A))$, while $T_{A,i}$ and μ_i are defined as above – supported in the solid torus defining σ_i.

Although it is not really necessary for our purposes here, it will be convenient to assume that the solid tori defining σ_i are disjoint, so make up a 'link' in the 3-manifold. We may then simplify our notation by writing T_A for $T_{A,i}$, with the obvious meaning. The solutions of Equation 5.13 have a simple geometric interpretation, due to Floer, which is worth mentioning here although it will not play any role in our work. (See [8] for a more detailed account, in a case in which this interpretation is important.) If the solid tori are disjoint then Equation 5.13 and the fact that T_A is covariant constant in the loop direction imply that the connection A is reducible over each solid torus. Moreover for each torus the holonomies around all the loops γ_z are the same, with trace τ_i say. Likewise the holonomies around all meridians are the same, with trace θ_i say. The content of the equation is the condition that

$$\theta_i = \partial_i \eta(\tau_1, \ldots, \tau_N). \tag{5.14}$$

This is an equation which only involves the flat connection obtained by restricting A to the complement in Y of all the solid tori (since we can push the loop defining τ_i to the boundary). Conversely, if we are given such a flat connection one may construct a solution of Equation 5.13. In sum, just as the critical points of the original Chern–Simons functional correspond to representations of the fundamental group of Y, the critical points of the perturbed functional correspond to representations of the fundamental group of the link complement in Y, satisfying the boundary condition Equation 5.14.

The perturbed instanton equations on the cylinder, corresponding to the gradient of the perturbed functional can be written either in '3+1 notation' as

$$\frac{\partial A}{\partial t} = *_3 \left(F_A - \sum \partial_i \eta \; \mu_i \; T_A \right),$$

or in 4-dimensional notation as

$$F^+_{\mathbf{A}} = \sum \partial_i \eta \; \mu_i^+ \; T_{\mathbf{A}}. \tag{5.15}$$

Here $T_{\mathbf{A}}$ is defined in the obvious way and μ^+ is the self-dual 2-form, supported in a copy of $S^1 \times D^2 \times \mathbf{R}$, obtained by pulling back μ and taking the self-dual part. Now suppose that we have a 4-manifold X with a tubular end modelled on $Y \times \mathbf{R}$. We can introduce a cut-off function ψ supported on the cylindrical end and equal to 1 outside a compact set. Then we may consider solutions of the equation over the complete 4-manifold X:

$$F^+_{\mathbf{A}} = \psi \sum (\partial_i \eta) \; \mu_i^+ \; T_{\mathbf{A}}, \tag{5.16}$$

which obviously reduces to Equation 5.15 over the end. Similarly, if we have a pair of perturbations we can write down an equation over the infinite tube $Y \times \mathbf{R}$ which agrees with one perturbed equation for t large and positive and with the other for t large and negative.

With this differential geometry in place we can now turn to discuss the analysis. The holonomy around a circle is obtained by solving an ODE of the shape

$$\frac{du}{ds} = uA,$$

and if A is square integrable on the circle the solution u will lie in the Sobolev space L^2_1 and is therefore continuous. Moreover the value of the solution u regarded as a function of $A \in L^2$ will vary continuously. The restriction of an L^2_2 function in four dimensions will be in $L^2_{1/2}$ on a 1-dimensional submanifold, so certainly in L^2. This means that the whole set-up will behave well provided we work with connections in four dimensions of class at least L^2_2. We have perturbed linearised operators $L_{A,\underline{\eta}}$ which differ from L_A by a compact perturbation (in fact a pseudo-differential operator of order $-\infty$), so the spectral theory and linear analysis go through without change. The crucial issue is compactness. First we have a local discussion over a 4-ball. The perturbation terms $\partial_i \eta \; \mu_i \; T_A$ are obviously universally *bounded* in L^∞ for all connections. Uhlenbeck's theorem tells us that for suitable ϵ any sequence of connections over the ball with $\|F\|_{L^2} < \epsilon$ and $\|F^+\|_{L^\infty} \leq C$ has a subsequence converging (after gauge transformation) weakly in L^p_1 for all p. Second, extend to the subsets carrying the perturbation. The arguments for gluing gauge transformations work equally well with weak L^p_1 convergence, once $p > 2$. Thus if we have a sequence of solutions of Equation 5.15 over the tube we get weak L^p_1 convergence of a subsequence over products $N \times (a, b) \subset Y \times \mathbf{R}$ of a solid torus N defining the perturbation with a small interval (a, b) in the $\mathbf{R}$-variable, provided we can find a cover of this set by 4-balls over which the curvature of all the connections is small in L^2. Then restriction theorems for Sobolev spaces give strong L^2 convergence over the loops, once $p > 3$. In this way we can show that a subsequence converges weakly in L^p_1 over $N \times (a, b)$ to an L^p_1 solution of the Equation 5.15. From there it is fairly routine bootstrapping to show that the limit and convergence are both actually in C^∞.

Next we have to discuss the bubbling phenomenon. Again, our starting point is the fact that F^+ is universally bounded in L^∞ for any solution of Equation 5.15. The proof of the removable singularities theorem given in [11], [24] (or indeed, of the decay results in Chapter 4 above) extends easily to show that any connection over a punctured 4-ball with F^+ bounded in L^∞ and with $\|F\|_{L^2} < \infty$ has bounded curvature and may be represented over the entire 4-ball by a connection matrix in L^p_1, for all p. From this it follows, much as above, that if Z is a finite subset of $N \times (a, b)$ and A is a smooth connection over $(N \times (a, b)) \setminus Z$ whose curvature has finite L^2 norm and which satisfies Equation 5.15 almost everywhere, then A is gauge-equivalent to a smooth solution over $N \times (a, b)$. Similarly for the variants of the equation such as Equation 5.16.

Finally, we consider the global picture. It is obviously not appropriate to study solutions of the perturbed instanton equations over the entire cylinder with curvature in L^2, since the 'stationary solutions' obtained from the critical points will not satisfy this condition. Instead we study the solutions of the perturbed Equations 5.15 with

$$\int \left| F^- - \sum_i (\partial_i \eta) \; \mu_i^- \; T_{\mathbf{A}} \right|^2 < \infty.$$

In $(3+1)$-dimensional notation this is just the condition that

$$\int \left| \frac{dA}{dt} \right|^2 < \infty.$$

We may then take the whole discussion of Chapters 2, 3, 4 to the perturbed situation, systematically replacing the Chern–Simons functional by $\Phi + \underline{\eta}$. We get control of the connections over bands in the end as before using the universal L^∞ bound on the perturbation. The final conclusion is that the perturbed equations enjoy exactly the same analytical properties, including the partial compactness, as the original equations.

5.5.1 Transversality arguments

We will now justify our choice of perturbation by showing that these allow us to achieve the general position conditions assumed before. We begin with an elementary result, which we state in more generality than we need.

Lemma 5.13 *For $r \geq 1$ let $g_1, \ldots, g_M$ and $g_1', \ldots, g_M'$ be elements of $U(r)$ and suppose that for all words W*

$$\mathrm{Tr}(W(g_1, \ldots, g_M)) = \mathrm{Tr}(W(g_1', \ldots, g_M')).$$

Then there is a $u \in U(r)$ such that $g_i' = u g_i u^{-1}$ for each i.

Here a 'word' means, as usual, a formal expression in non-commuting variables. Thus another way of stating the hypothesis of the lemma is that if ρ, ρ' are two homomorphisms from the free group Fr_M on M symbols to $SU(r)$ and if $\mathrm{Tr}(\rho(X)) = \mathrm{Tr}(\rho'(X))$ for all X in Fr_M then there is a u such that $\rho'(X) = u\rho(X)u^{-1}$ for all X. Said differently again, the assertion is that two finite-dimensional unitary representations of Fr_M with the same character are conjugate. For the proof, we let $\mathfrak{I}(\rho, \rho')$ be the image of the homomorphism (ρ, ρ') from Fr_M to $SU(r) \times SU(r)$. This is a subgroup of $SU(r)$ and by assumption for any element (v, v') of $\mathfrak{I}(\rho, \rho')$ we have $\mathrm{Tr}(v) = \mathrm{Tr}(v')$. Let H be the closure of $\mathfrak{I}(\rho, \rho')$ in $SU(r) \times SU(r)$. By a fundamental theorem from Lie group theory this is a Lie subgroup, and plainly $\mathrm{Tr}(v) = \mathrm{Tr}(v')$ for any $(v, v') \in H$. In other words the two representations of H obtained by projecting to the two factors in $SU(r) \times SU(r)$ have the same character. But, by another basic theorem, we know that a representation of a *compact* group is determined up to conjugacy by its character and our result follows.

Now by a straightforward compactness argument we can deduce that, for fixed r, M, there is a finite number of words $W_1, \dots, W_N$ such that if

$$\operatorname{Tr}(W_\alpha(g_1, \dots, g_M)) = \operatorname{Tr}(W_\alpha(g_1', \dots, g_M'))$$

for all $\alpha = 1, \dots, N$, then the g_i, g_i' are simultaneously conjugate, as above. In the case most relevant to us, when the elements lie in $SU(2)$, it almost suffices to take the $M + M(M-1)/2$ words

$$x_i,\ i = 1, \dots, M, \quad x_i x_j^{-1},\ i, j = 1, \dots, M,\ i < j. \tag{5.17}$$

It is an easy and enjoyable exercise to check that if the traces of these words, evaluated on two systems g_i, g_i' in $SU(2)$, are equal then either the g_i, g_i' are simultaneously conjugate as above or there is a u such that $g_i' = u g_i^{-1} u^{-1}$. Thus we need some extra words, for example the commutators, to rule out the latter alternative.

Consider the product $SU(r)^M$ and the action of $SU(r)$ on this set by simultaneous conjugation. Write $SU(r)^M/SU(r)$ for the quotient space. For any word W the trace defines a function $\chi_W : SU(r)^M/SU(r) \to \mathbf{R}$. Thus if we fix a set of N words W_α as above we get a map

$$\underline{\chi} : SU(r)^M/SU(r) \to \mathbf{R}^N.$$

The content of the discussion above is that we can choose words so that $\underline{\chi}$ is an *embedding* of the quotient space in $\mathbf{R}^N$. Now let $(SU(r)^M)_*$ denote the subset of $SU(r)^M$ of points whose stabiliser under the action consists only of the centre of $SU(r)$. The quotient of this subset is then a manifold.

Lemma 5.14 *We may choose words W_α such that $\underline{\chi}$ is an immersion of $(SU(r))^M_*/SU(r)$ in $\mathbf{R}^M$.*

For simplicity we do this in the case of $SU(2)$, and we consider the set of $N_0 = M + M(M-1)/2$ words given in Formula 5.17 above. (Obviously the immersive condition is preserved if we extend our set of words.) We identify the tangent space to $SU(2)^M$ at a point $(g_1, \dots, g_M)$ with the product of M copies of the Lie algebra, by left translation. The derivative of $\underline{\chi}$ at this point is then given by the linear map $L : \mathfrak{su}(2)^M \to \mathbf{R}^{N_0}$;

$$L(\xi_1, \dots, \xi_M) = (\langle \xi_i, \gamma_i \rangle; \langle \xi_i - \xi_j, \gamma_{ij} \rangle).$$

Here $\langle\,,\rangle$ denotes the standard inner product on $\mathfrak{su}(2) = \mathbf{R}^3$ and

$$\gamma_i = g_i - \tfrac{1}{2}\operatorname{Tr}(g_i)1, \quad \gamma_{ij} = g_i g_j^{-1} - \tfrac{1}{2}\operatorname{Tr}(g_i g_j^{-1})1.$$

(The expression appearing here is just the projection π from $SU(2)$ to its Lie algebra considered in Equation 5.12 above.) It is again an exercise in linear algebra, using induction on M, to show that, provided the g_i do not lie in a single one-parameter subgroup, this map L has a 3-dimensional kernel, the set of $(\xi_1, \ldots, \xi_M)$ with

$$\xi_i = \eta - g_i \eta g_i^{-1}$$

for $\eta \in \mathfrak{su}(2)$. Of course the problem here is just an infinitesimal version of the one considered previously.

To sum up, we have shown that we may choose words so that the map $\underline{\chi}$ gives an embedding of $SU(2)^r/SU(2)$ in Euclidean space, and is an immersion on the manifold subset $(SU(2)^r)_*/SU(2)$.

Return now to the space of $SU(2)$ connections over a homology 3-sphere Y. Let $K \subset \mathcal{B}_Y^*$ be a compact subset and $\tilde{K} \to K$ be a compact subset of the tangent bundle of $\mathcal{B}_Y^*$, restricted to K. For example, if K is a submanifold, $\tilde{K}$ could be the unit sphere bundle in TK. It is a basic fact that a connection is determined, up to gauge equivalence, by its holonomy. The infinitesimal version of this is the fact that for any tangent vector a to $\mathcal{B}_Y^*$ at a connection A we may choose a finite set of loops δ_i, $i = 1, \ldots, M$, in Y, with a common base point, such that the derivatives of the holonomy maps around δ_i map a to a non-zero tangent vector of $(SU(2)^M)_*/SU(2)$. By compactness, we may choose these loops such that this derivative maps the fibres of $\tilde{K} \to K$ injectively to the tangent spaces of $(SU(2)^M)_*/SU(2)$. Fix an map $\underline{\chi}$ as above, defined by words W_α. For each word W_α we form a loop $\tilde{\gamma}_\alpha$ out of the δ_i, using the usual product operation. Then the traces of the holonomies around the loops $\tilde{\gamma}_\alpha$ map the fibres of $\tilde{K} \to K$ injectively into the tangent space of $\mathbf{R}^N$. The loops $\tilde{\gamma}_\alpha$ cannot be disjoint and embedded but we may approximate them arbitrarily closely by disjoint embedded loops γ_α in Y and it is clear that, for a good enough approximation, the traces of the holonomies around the γ_α will have the same property. Likewise, if we perform the averaging construction of Equation 5.10 with a sufficiently concentrated 2-form μ we see that we can find functions σ_α – for $\alpha = 1, \ldots, N$ – on the space of connections, whose derivatives give fibrewise embeddings.

Now suppose that K has local 'finite-dimensional models', as we have seen holds true when K is the space $\mathcal{R}_Y^*$ of flat connections. Thus we suppose that in a neighbourhood of any point K is contained in a finite-dimensional submanifold. It follows easily then from the discussion above (adjoining the unit spheres in the tangent spaces of these finite-dimensional manifolds to $\tilde{K}$) that we can choose $\underline{\sigma} = (\sigma_1, \ldots, \sigma_N)$ in such a way that $\underline{\sigma}$ gives an embedding of K in $\mathbf{R}^N$ and the derivative $D(\underline{\sigma})$ of $\underline{\sigma}$ gives an embedding of $\tilde{K}$ in the tangent space of $\mathbf{R}^N$. (That is, we first choose our loops to get an embedding in $(SU(2)^M)_*/SU(2)$ and then apply the Lemma to embed in $\mathbf{R}^N$.)

The work that remains now is fairly standard differential topology. Recall that $\mathcal{R}_Y^*$ denotes the set of equivalence classes of irreducible flat connections, the critical points of the unperturbed problem. Let $\tilde{\mathcal{R}}_Y^*$ be the set of equivalence classes of pairs (A, a) where $[A] \in \mathcal{R}_Y^*$ and $a \in H_A^1$ with $\|a\| = 1$. (This can be thought of as the set of unit vectors in the Zariski tangent space of $\mathcal{R}_Y^*$.)

Proposition 5.15 *Suppose $\underline{\sigma}$ is chosen so that $D(\underline{\sigma})$ gives an embedding of $\tilde{\mathcal{R}}_Y^*$ in $T\mathbf{R}^N$. Then we may find arbitrarily small $\underline{\epsilon} \in \mathbf{R}^N$ such that if $\eta(x_1, \ldots, x_N) = \sum \epsilon_i x_i$ the critical points of the functional $\Phi + \underline{\eta} : \mathcal{B}^* \to \mathbf{R}/\mathbf{Z}$ are non-degenerate.*

To prove this we pull back the cotangent bundle of $\mathcal{B}_Y^*$ to get a bundle, $\mathcal{E}$ say, over $\mathcal{B}_Y^* \times \mathbf{R}^N$. Thinking of ϵ_i as co-ordinates on $\mathbf{R}^N$, the expression $\Phi + \sum \epsilon_i d\sigma_i$ defines a section, $\underline{\Phi}$ say, of $\mathcal{E}$. The hypothesis says that the derivative of $\underline{\Phi}$ is surjective at all points in $\mathcal{R}_Y^* \times \{0\}$. So, by the implicit function theorem, the zero set $\underline{Z}$ of $\underline{\Phi}$ is a smooth N-dimensional submanifold in a neighbourhood of this set. Now consider the projection π from $\underline{Z}$ to $\mathbf{R}^N$. By Sard's theorem there are arbitrarily small $\underline{\epsilon}$ which are regular values of π, and any such regular value satisfies the condition stated in the Proposition.

Notice that, since the zero set $\mathcal{R}_{\mathcal{Y}}^*$ is compact, the set of parameters ϵ which satisfy the condition of Proposition 5.15 is actually open and dense.

We now move on to the gradient lines. We should first fix a perturbation as in Proposition 5.15 to make the critical points of the perturbed functional non-degenerate then we have a good Fredholm theory for solutions of the deformed instanton equation as discussed above. The remaining issue is to arrange, by making a further small perturbation, that all the moduli spaces of deformed instantons are cut out transversally. To simplify our notation let us suppose that in fact the original Chern–Simons functional has non-degenerate critical points and focus on the further deformation which needs to be made.

Suppose then that $\underline{A}$ is an instanton over the tube and the cokernel $H^2_{\underline{A}}$ is non-zero. Recall that if we view $\underline{A}$ as a one-parameter family of connections over Y, the vector space $H^2_{\underline{A}}$ can be identified with the time-dependent bundle-valued 1-forms ϕ over Y satisfying the equation

$$\frac{d\phi}{dt} = - *_3 d_A *_3 \phi,$$

which decay as $t \to \pm\infty$.

Lemma 5.16 *For each t the L^2 inner product over Y,*

$$\left\langle \phi, \frac{dA}{dt} \right\rangle$$

is zero.

To prove this we recall that the instanton condition is $\frac{dA}{dt} = *_3 F_A$ and differentiate with respect to t:

$$\begin{aligned} \frac{d}{dt}\left\langle \phi, \frac{dA}{dt} \right\rangle &= \left\langle \frac{d\phi}{dt}, *_3 F_A \right\rangle + \left\langle \phi, *_3 \frac{dF}{dt} \right\rangle \\ &= -\langle *_3 d_A *_3 \phi, *_3 F_A \rangle + \langle \phi, *_3 d_A *_3 F_A \rangle = 0. \end{aligned}$$

Thus the inner product is constant in t, but since all terms decay at $\pm\infty$ the constant must be zero.

(There is a more conceptual proof of the lemma involving reparametrisations of the gradient line: in particular the analogous result is true for any gradient equation.)

Now a variant of the 'unique continuation' result of [17][Chapter 4] shows that the path A_t, viewed as a map from $\mathbf{R}$ to $\mathcal{B}_Y$, is one-to-one and maps into the set of irreducible connections. Since the A_t converge to critical points at $\pm\infty$ the image is contained in a compact set in $\mathcal{B}_Y$. We may thus choose our map $\underline{\sigma}$ as above so that the composite $\underline{\sigma}(A_t)$ defines an embedded path ν in $\mathbf{R}^N$. Fix some $t_0 \in \mathbf{R}$ such that ϕ_{t_0} does not vanish (in fact, by a unique continuation result, any t_0 will do). By the immersive condition and the Lemma we may suppose $\underline{\sigma}$ is chosen such that the images under $D\underline{\sigma}$ of $\frac{dA}{dt}$ and ϕ, evaluated at t_0, are a pair of linearly independent vectors, p, q say, in $\mathbf{R}^N$. Of course X is just the velocity vector of the path ν at the parameter value t_0. Now we choose a small neighbourhood U of $\nu(t_0)$ in $\mathbf{R}^N$ such that the intersection $\nu^{-1}(U)$ of the path ν with U is a small interval about t_0. It is clear then that we can find a function η on $\mathbf{R}^N$ supported in the small set U such that

$(d\eta)_{\nu(t_0)}(q) > 0$ and moreover $(d\eta)_{\nu(t)}(q_t) \geq 0$ for all t, where

$$q_t = (D\underline{\sigma})_{A_t}(\phi_t).$$

This means that, writing $\underline{\eta} = \eta \circ \underline{\sigma}$ as before,

$$\int_{-\infty}^{\infty} d\underline{\eta}(\phi_t)\, dt > 0.$$

(Here ϕ_t is regarded as a tangent vector to $\mathcal{B}_Y^*$ at A_t.)

Now let K be any compact subset of a moduli space of instantons over the tube. Evaluating at some t_0, this maps to a compact set in $\mathcal{B}_Y^*$. Similarly the set of unit vectors in the cokernel spaces $H^2_{\underline{A}}$ for $\underline{A} \in K$ forms a compact set of tangent vectors. By using the construction above, and a covering argument, we can find a $\underline{\sigma}$ and a finite set of functions $\eta_1, \eta_2, \ldots, \eta_r$ on $\mathbf{R}^N$ such that for any $\underline{A} \in K$ and $\phi \in H^2_{\underline{A}}$ there is an η_α with

$$\int_{-\infty}^{\infty} d\eta_\alpha(\phi_t)\, dt \neq 0.$$

We can now use the same standard argument as in the proof of Proposition 5.15. We consider the family of perturbed instanton equations parametrised by $\underline{\delta} \in \mathbf{R}^r$:

$$\frac{dA}{dt} = *F_A + \mathrm{grad}\, \underline{\eta}^{\underline{\delta}}, \tag{5.18}$$

where $\underline{\eta}^{\underline{\delta}}$ is the function

$$\underline{\eta}^{\underline{\delta}} = \sum_{\alpha=1}^{r} \delta_\alpha \underline{\eta}_\alpha$$

on $\mathcal{B}_Y$. We regard Equation 5.18 as defining a section of a suitable bundle over the product of the space of connections over the cylinder with the parameter space $\mathbf{R}^r$. The condition above asserts that this section has transverse zeros near $K \times \{0\}$, and we then apply Sard's theorem to the projection map to $\mathbf{R}^r$. We obtain a perturbation $\underline{\eta}^{\underline{\delta}}$ such that the perturbed instanton moduli space, in a neighbourhood of K, is cut out transversally.

Our discussion of the gluing theory for instantons on the cylinder shows that if a sequence of instantons $\underline{A}_i$ is chain-convergent and if the H^2 spaces vanish for all the instantons in the limiting chain, then $H^2_{\underline{A}_i} = 0$ for large i. Similarly for weak chain convergence in which bubbling off may occur. The corresponding results are true for the perturbed instanton equations. This means that we can extend our transversality results from the compact sets considered above to all instanton moduli spaces, using induction on the energy level. To sum up then we have

Proposition 5.17 *There are arbitrarily small perturbations $\underline{\eta}$ such that all the critical points of $\Phi + \underline{\eta}$ are non-degenerate and all perturbed instanton moduli spaces are regular.*

With all this discussion in place we are now able to define the Floer groups for a general homology 3-sphere. We choose a suitable small perturbation $\underline{\eta}$ and apply the construction of Sections 5.2 and 5.4 to the perturbed functional. The argument of Section 5.3, applied to suitable deformed instanton equations, shows that the Floer groups are independent, up to canonical isomorphism, of the perturbation.

There is one important point to notice here, involving the reducible connection θ. We know that this is a non-degenerate critical point of the original Chern–Simons functional Φ since $H^1_\theta = H^1(Y; \mathbf{R}) \otimes \mathfrak{su}(2) = 0$. Thus the same will be true for small enough perturbations $\Phi + \underline{\eta}$. More precisely, we want θ to be a non-degenerate critical point of the functions $\Phi + s\underline{\eta}$ for all $s \in [0, 1]$. If we have a pair of such perturbations the index for the perturbed instanton equation over the cylinder with limits θ at $\pm\infty$ (in the trivial homotopy class) is -3, as before. This means that the only perturbed instanton of this kind is the trivial connection over the cylinder, which is the essential fact used in Section 5.3. Another point to note is that we can use the perturbations of the instanton equation discussed above to get around a problem with the flat connections over 4-manifolds. Recall that this is the case not covered by the Freed–Uhlenbeck generic metrics theorem: one obviously cannot make the moduli space of flat instantons regular by changing the metric. However, one can achieve regular moduli spaces by making a small perturbation of the kind discussed here.

5.6 $U(2)$ and $SO(3)$ connections

Throughout our work so far we have concentrated on $SU(2)$ connections. We will now consider a straightforward extension of the theory, involving the structure groups $U(2)$ and $SO(3)$. Consider first a $U(2)$ bundle E over a manifold V (it will be convenient to work with complex vector bundles here). The bundle $\mathfrak{g}_E$ of trace-free, skew adjoint, automorphisms of E is a real vector bundle with structure group $SO(3)$. The characteristic classes of these bundles are related by

$$w_2(\mathfrak{g}_E) = c_1(E) \bmod 2, \quad p_1(\mathfrak{g}_E) = c_1(E)^2 - 4c_2.$$

An $SO(3)$ bundle W lifts to $U(2)$, i.e. arises as $\mathfrak{g}_E$ for some E, if and only if the Stiefel–Whitney class $w_2(W) \in H^2(V; \mathbf{Z}/2)$ can be lifted to an integral class. A connection on E induces connections of $\mathfrak{g}_E$ and on the line bundle $\Lambda^2 E$; conversely it is easy to see that connections on these two bundles determine a unique connection on E. The advantage of working with $U(2)$ bundles is that the discussion of 'reducible connections' is straightforward. A $U(2)$ connection has a non-trivial stabiliser under the action of the gauge group (that is, larger than the scalars) if and only if it is compatible with splitting $E = L_1 \oplus L_2$. In the case of $SO(3)$ on the other hand there are a number of different possibilities; see [8]. Fix a connection α on the line bundle $\Lambda^2 E$ and let $\mathcal{A}_E$ be the set of connections on E compatible with α. Let $\mathcal{G}_E$ be the unitary automorphisms of E of determinant 1; then the stabiliser of a generic connection in $\mathcal{A}$ is ± 1, just as in the $SU(2)$ picture, and the reducible connections, with non-trivial stabiliser, arise from splittings as above. There is a natural map from $\mathcal{A}_E/\mathcal{G}_E$ to the space $\mathcal{A}_{\mathfrak{g}_E}/\mathcal{G}_{\mathfrak{g}_E}$ of equivalence classes of connections on the $SO(3)$ bundle $\mathfrak{g}_E$ and this map is finite-to-one. In fact there is an action of the group $H^1(V; \mathbf{Z}/2)$ on $\mathcal{A}_E/\mathcal{G}_E$ and $\mathcal{A}_{\mathfrak{g}_E}/\mathcal{G}_{\mathfrak{g}_E}$ is the quotient space. To see this action, think of an element χ of $H^1(V; \mathbf{Z}/2)$ as a homomorphism from π_V to $\pm 1 \subset S^1$. As such it defines a flat complex line bundle L_χ such that L_χ^2 is trivial (as a line bundle-with-connection). This mean that $E \otimes L_\chi$ is isomorphic to E and if A is a connection in $\mathcal{A}_E$ the connection $A \otimes \chi$ induced by A and χ on $E \otimes L_\chi$ defines the same connection α on $\Lambda^2 E = \Lambda^2(E \otimes L_\chi$. Thus the action of χ is given by

$$\chi([A]) = [A \otimes \chi].$$

It is clear that A and $A \otimes \chi$ define the same connection on $\mathfrak{g}_E$ and that an $SO(3)$ connection arises from a unique $U(2)$ connection, up to this action. In this framework, the possible complications of the stabilisers in the $SO(3)$ gauge group arise as the stabilisers of the $H^1(V; \mathbf{Z}/2)$ action on $\mathcal{A}_E$.

Now let Y be a 3-manifold and X be an oriented Riemannian 4-manifold. We study $U(2)$ connections over X and Y which induce flat connections or instantons respectively on the associated $SO(3)$ bundles, with some arbitrary, fixed, connection on Λ^2, as above. The connection on Λ^2 plays no real geometrical role – it is obvious that different choices give equivalent moduli spaces – but this set-up gets around some technical difficulties which arise if one tries to work directly with $SO(3)$ connections. For simplicity of language we will still refer to these $U(2)$ connections as 'flat' connections and 'instantons' respectively, with the understanding that they are really *projectively* flat connections/instantons, with fixed determinant.

We say that a $U(2)$ bundle E over a manifold V is 'admissible' if $c_1(E)$ defines an *odd* element of the free abelian group $H^2(V;\mathbf{Z})/\mathrm{Torsion}$. The importance of this is that an admissible bundle E cannot support any flat connection. For the hypothesis implies that there is a surface $\Sigma \subset V$ such that E has odd degree over Σ. But if E is reducible there is a splitting $E = L_1 \oplus L_2$ and the 'flat' condition implies that the curvatures of L_1 and L_2 are equal. The Chern–Weil theory tells us that the degrees of L_1 and L_2 over Σ are equal, so the degree of E is even; a contradiction. In four dimensions this means that the theory of $U(2)$ moduli spaces on an admissible bundle E follows just the same lines as the $SU(2)$ case summarised in Chapter 2. If $b^+ > 0$ there are *no* reducible solutions for generic metrics and if $b^+ > 1$ there are no reducibles in generic one-parameter families. The moduli space dimension is given by

$$\dim M_E = 2(c_1(E)^2 - 4c_2(E)) - 3(1 - b_1 + b_+). \tag{5.19}$$

The compactification of the moduli space involves bundles with the same c_1 and lower values of c_2. Thus we get a theory for each class $c \in H^2(X;\mathbf{Z})$ which is odd in $H^2/\mathrm{Torsion}$. Classes c, c' which differ by an even element 2λ give equivalent theories: we just replace E by $E \otimes L$ where $c_1(L) = \lambda$. The moduli spaces are all canonically oriented by a choice of 'homology orientation' of X, just as in the $SU(2)$ case; however, the equivalence between the theories for E and $E \otimes L$ changes the orientation according to the parity of $c_1(L)^2$.

We now turn to the Floer groups. Let Y be an oriented Riemannian 3-manifold and E be an admissible bundle over Y. (Note that this means that the first Betti number of Y is non-zero). The Chern–Simons functional is defined as a map from $\mathcal{A}_E/\mathcal{G}_E$ to $\mathbf{R}/\mathbf{Z}$, much as before, and its critical points are the flat connections. The holonomy of these flat connections takes values in copies of $SU(2)$ (since the determinant is fixed), so the discussion of perturbations from the previous Section can be taken over with only minor changes. A loop in $\mathcal{A}_E/\mathcal{G}_E$ gives a connection on a bundle $\mathbf{E}$ over $Y \times S^1$. The fact that the connections are fixed on $\Lambda^2 E$ means that the first Chern class of $\mathbf{E}$ is pulled back from that of E under the projection from $Y \times S^1$ to Y: there is no component in

$$H^1(Y) \otimes H^1(S^1) = H^1(Y) \subset H^2(Y \times S^2). \tag{5.20}$$

Thus the index associated to $\mathbf{E}$ by Equation 5.19 is 0 modulo 8 and this means that we get a $(\mathbf{Z}/8)$-relative index $\delta(\rho, \sigma)$, as before, and hence a graded chain group. An important difference is that the grading is, initially, only an *affine* $(\mathbf{Z}/8)$-grading; that is we need to make an arbitrary choice of the term of degree 0 since we do not have the trivial connection to compare with, as in the $SU(2)$ case. The definition of the Floer differential, and the verification that $\partial^2 = 0$, go through without essential change.

The proof that the Floer groups are independent of choices brings in some new features. As before, the crucial thing is to show that a cobordism X from Y_0 to Y_1 induces a map on the Floer groups. More precisely, now we should suppose that we have a bundle $\mathbf{E}$ over X which restricts to the chosen bundles $E_i \to Y_i$ over the ends, and fix a connection α on $\Lambda^2\mathbf{E}$, agreeing with the given ones over the ends. Now recall that the group $H^1(Y; \mathbf{Z}/2)$ acts on the space $\mathcal{A}_E/\mathcal{G}_E$. The action preserves the critical points of the Chern–Simons functional on Y. One might expect that the Floer complex can also be chosen to be invariant under the action, which would lead to an $H^1(Y; \mathbf{Z}/2)$-action on the Floer homology. However, there are possible obstructions to making the desired perturbation – to achieve transversality for instanton moduli spaces over the tube – invariant under the action. So instead we proceed as follows. We restrict attention to the subgroup $G \subset H^1(Y; \mathbf{Z}/2)$ of classes which lift to integer co-efficients, i.e. the kernel of the Bockstein map from $H^1(Y; \mathbf{Z}/2)$ to $H^2(Y; \mathbf{Z})$. For a class $\chi \in G$ the line bundle L_χ is trivial. Explicitly, L_χ can be constructed as the trivial bundle-with-connection form ia where $da = 0$ and

$$\int_\gamma a = \pi \hat{\chi}(\gamma),$$

and $\hat{\chi}$ is an integral lift of χ. Thus there is a connection α_X on the trivial line bundle L over $Y \times [0,1]$ which is trivial over $Y \times \{0\}$ and equal to ia over $Y \times \{1\}$. Taking this as our fixed connection on Λ^2, we get an induced map on the Floer chains defined by the trivial cobordism but this non-trivial α_X, in the familiar way. The conclusion is that there is indeed an action of G on the Floer groups, though it is a subtle question whether this can be induced by an action on a suitably perturbed complex. If we are given a cobordism X from Y_0 to Y_1 we need to specify both the topological type of the bundle $\mathbf{E}$ over X and also a choice of the connection on the determinant (modulo a certain equivalence relation). Different choices of the latter give different maps on the Floer groups – differing by these group actions. In the proof that the Floer groups are independent of choices we are concerned with a cobordism X which is topologically a product $Y \times [0,1]$ and we fix the connection on Λ^2 to be that pulled back from the projection to Y and this fixes the ambiguity. Then the proof of invariance goes through as before.

The action of $G \subset H^2(Y;\mathbf{Z}/2)$ on the Floer groups does not in general preserve the grading. To see this, let $\mathbf{A}$ be a connection on the bundle $\pi^*(E)$ over $Y \times \mathbf{R}$ which restricts to A_0 over $Y \times \{0\}$, to a connection isomorphic to $A_0 \otimes \chi$ over $Y \times \{1\}$, and with fixed determinant equal to $\pi^*(\alpha_Y)$ over all of $Y \times [0,1]$. According to the general procedure of Chapter 2, the shift in the grading induced by χ is given by the index of the D operator on an $SO(3)$ bundle over $Y \times S^1$ obtained by gluing $\pi^*(\mathfrak{g}_E)$ over the two ends of $Y \times [0,1]$. If we assume as before that χ is the reduction of an integral class $\hat{\chi}$, then we can obtain this $SO(3)$ bundle from a $U(2)$ bundle $\mathbf{E}$ over $Y \times S^1$ as follows. The connection $A_0 \otimes \chi$ is isomorphic but not equal to a connection on E with determinant α_Y; that is, there is a unitary automorphism g of E such that $g^*(A_0 \otimes \chi) - A_0$. Then $\mathbf{E}$ is obtained by gluing $\pi^*(E)$ by the automorphism g over the ends. Now, as before, we can write $A_0 \otimes \chi$ as $A_0 + ia1$. Let $h : Y \to S^1$ be the determinant of g. Then we have $dhh^{-1} = 2ia$, so h defines the class $\hat{\chi}$ in $H^1(Y;\mathbf{Z})$. It follows then that the $U(2)$ bundle $\mathbf{E}$ has first Chern class

$$c_1(E) + \hat{\chi}\theta \in H^2(Y) \oplus H^1(Y) \otimes H^1(S^1) = H^2(Y),$$

where θ is the fundamental class of S^1. Hence

$$\langle c_1(\mathbf{E})^2, [Y \times S^1] \rangle = 2\langle \hat{\chi} c_1(E), [Y] \rangle.$$

It follows then from Equation 5.19 that the index of D on the associated $SO(3)$ bundle is equal to $4\chi c_1(E)$ modulo 8, so the action of G preserves the grading modulo 4. (One can extend, at least in part, this discussion of the group action from the group G to all of $H^1(Y;\mathbf{Z}/2)$, by using $SO(3)$ bundles rather than $U(2)$ bundles. However, there are difficulties in fixing the orientations in that setting.)

Finally we return to the question of the grading in this $U(2)$ set-up. Here we follow the discussion of Froyshov in [25]. Suppose ρ is a flat connection on an admissible bundle E over Y. We can find a 4-manifold X_1 with boundary Y and an adapted bundle E_1 over X extending ρ.

Reversing orientation, we can find a 4-manifold Z with boundary $\overline{Y}$ and a similar extension E_Z over Z. Then by gluing we can form a closed 4-manifold $X = X_1 \cup_Y Z$ and a bundle $\mathbf{E}$ over Z. The dimension equation Formula 5.19 and our gluing formula tell us that

$$\operatorname{ind} E_1 + \operatorname{ind} E_Z = -3(1 - b_1(X) + b_+(X)) \mod 2,$$

since the contribution $2(4c_2 - c_1^2)$ is plainly even. On the other hand we know that

$$(1 - b_1(X) + b_+(X)) = \operatorname{ind}^-(X_1) + \operatorname{ind}^-(Z) - (1 + b_1(Y).$$

So

$$\operatorname{ind} E_1 - 3 \operatorname{ind}_-(X_1) = -\operatorname{ind} E_Z + 3 \operatorname{ind}_-(Z) - 3(1 + b_1(Y)) \mod 2,$$

and the right hand side is independent of the choice of X_1. So we conclude that, modulo 2, $\operatorname{ind} E_1 - 3 \operatorname{ind}_-(X_1)$ is an invariant of the flat connection ρ and this enables us to fix the grading of the Floer groups modulo 2. We define

$$\delta(\rho) = \operatorname{ind} E_1 - 3 \operatorname{ind}_-(X_1) \mod 2,$$

to fix the mod 2 grading. (Recall that

$$\operatorname{ind}_-(X_1) = b_1 - b_+,$$

where b_+ is the dimension of a maximal positive subspace for the intersection form on $\operatorname{Im} : H^2(X_1, Y) \to H^2(X_1)$.)

We can refine this discussion by fixing a *spin structure* on Y. Then we may take the 4-manifolds X_1, Z above to be spin 4-manifolds, inducing the given structure on the boundary. This means that X is a closed, spin, 4-manifold and so has even intersection form. Thus the term c_1^2 in Equation 5.19 is even and hence $2(4c_2 - c_1^2)$ is zero modulo 4. It follows that, defined in this way, $\delta(\rho)$ is well-defined modulo 4; hence we see that a spin structure on Y defines a definite $(\mathbf{Z}/4)$-grading on the Floer groups.

6

Floer homology and 4-manifold invariants

In this Chapter we discuss the relationship between the Floer homology groups and the invariants of 4-manifolds defined by Yang–Mills instantons. We have already noted in Chapter 1 that this relation has the general shape of a 'topological quantum field theory' (TQFT), and we have studied a prototype situation in Chapter 5, where we saw that an h-cobordism W between homology 3-spheres Y_0, Y_1 induces a map

$$\zeta_W : HF_*(Y_0) \to HF_*(Y_1). \tag{6.1}$$

Before beginning the technical development in earnest we digress to give some motivation for the general picture; motivation which is very important in understanding the significance of Floer's basic construction. In Section 6.3 we summarise the theory of instanton invariants of closed 4-manifolds and the vanishing theorem for the invariants of connected sums. (One could think of this as a rather simple prototype for the more general splittings encompassed by Floer's theory.)

6.1 The conceptual picture

We begin by contemplating the standard picture in finite-dimensional Morse Theory. That is, we consider a compact n-manifold B and a Morse function $f : B \to \mathbf{R}$. There are a finite number of critical points b_r of f, and the Hessian of f at b_r is a non-degenerate quadratic form on the tangent space of B. The *index* $i(b_r)$ is the dimension of a maximal negative subspace for the Hessian, the 'number of negative eigenvalues'. The fundamental assertion of Morse theory is that the homology of B may be computed from a chain complex

$$(C_*, \partial) \tag{6.2}$$

where the chain group C_λ is the free abelian group with generators the critical points b_r with index λ. (More precisely, to build in signs, we should take the generators to be pairs consisting of a critical point b_r and an orientation $\mathcal{O}$ of a maximal negative subspace for the Hessian at b_r, and identify $-(b_r, \mathcal{O})$ with $(b_r, \overline{\mathcal{O}})$, where $\overline{\mathcal{O}}$ is the opposite orientation.) There are many points of view on this: one way goes via a cell decomposition of B induced by the gradient flow $\phi_t : B \to B$; the one-parameter family of diffeomorphisms generated by the vector field $\mathrm{grad}\, f$ (defined using some fixed Riemannian metric on B). For any point x in B the gradient flow $\phi_t(x)$ converges to some critical point b_r of f – a zero of $\mathrm{grad}\, f$ – as $t \to \infty$, for a suitable $r = r(x)$. Typically this limit is a local maximum of f, that is, it has index n. More precisely, this is true for an open, dense set $B_n \subset B$. This subset B_n has one component for each critical point of index n. For large t the gradient flow contracts any given compact subset of one of these components into a small neighbourhood of the critical point, and one sees from this that each component is diffeomorphic to an open n-ball. Similarly, the set B_λ of points which flow as $t \to \infty$ to critical points of index λ is a union of open λ-balls, embedded in B. Thus we can think of building up the entire manifold B by starting with B_0 – the finite set of minima – and successively adding the higher-dimensional cells making up the B_λ for $\lambda > 0$, finishing off with the n-cells. The additional fact we need to know, for this procedure to give a cell decomposition, is that all cells are attached to cells of lower dimension, i.e. that

$$\overline{B_\lambda} \subset B_0 \cup \cdots \cup B_\lambda. \tag{6.3}$$

This is true for a generic choice of function f and Riemannian metric on B. To understand why, we introduce, for each critical point b_r, the ascending and descending manifolds U_r, V_r consisting of points which flow to b_r as $t \to +\infty, -\infty$ respectively. So U_r is just one of the cells considered in the decomposition $\{B_\lambda\}$ above, and V_r appears in just the same way if one replaces the function f by $-f$. Suppose b_s is another critical point, with $f(b_s) > f(b_r)$. Choose some level c which is a regular value of f and with $f(b_s) > c > f(b_r)$. Then $W = f^{-1}(c)$ is a smooth $(n-1)$-manifold. The descending set V_r meets W in an embedded sphere Σ_r of dimension $\dim V_r - 1 = n - i(r) - 1$ and symmetrically $U_s \cap W = \Sigma_s$ is an embedded sphere of dimension $i(s) - 1$. We assume that the set-up satisfies the *Smale condition*, that for any r, s these spheres meet transversally in W. In particular, if $\dim \Sigma_r + \dim \Sigma_s < \dim W = n - 1$, that is, if $i(r) \geq i(s)$, then the spheres are disjoint. Now from the definitions, the points of intersection of spheres Σ_r, Σ_s are in one-to-one correspondence with integral curves $x(t)$ of the vector field $\text{grad} f$ with $x(t) \to b_r$ as $t \to -\infty$ and $x(t) \to b_s$ as $t \to +\infty$, modulo the obvious change of parametrisation $t \mapsto t + c$. We will call such curves *gradient trajectories from b_r to b_s*. So, to sum up, we see that, if the Smale condition holds, there are no gradient trajectories between points b_r and b_s with $i(r) \geq i(s)$.

Now to establish Formula 6.3, we suppose that x is a point in the closure $\overline{B_\lambda}$. Thus there are a critical point b_s of index λ and a sequence x_i in $U_s \subset B$ converging to x. For each i there is an integral curve $x_i(t)$ for $t \in [0, \infty)$ with $x_i(0) = x_i$ and $x_i(t) \to b_s$ as $t \to \infty$. The key then is to analyse the sense in which these curves converge (possibly after taking subsequences). First, by considering the behaviour for large t, one shows that there is an increasing sequence $c_i \geq 0$ such that the curves $\overline{x_i}(t) = x_i(t+c_i)$ converge on compact subsets of $(-\overline{c}, \infty)$ where $\overline{c} = \lim c_i$. Now different cases arise. If $\overline{c}$ is finite, then the limit is an integral curve starting with x and converging to b_s, so in this case x is in U_s and is a point of B_λ. On the other hand if $\overline{c}$ is infinite the limit is a gradient trajectory from some other critical point to b_s. Arguing in a similar way, one finds a chain of critical points $b_{s_1}, b_{s_2}, \ldots, b_{s_k}$ say, with $b_{s_k} = b_s$, such that there are gradient trajectories joining consecutive pairs b_{s_i}, $b_{s_{i+1}}$ and that the original point x lies in U_{s_1}. Now the existence of the gradient trajectories and the result of the previous paragraph show that

$$i(s_1) < i(s_2) < \cdots < i(s_k) = \lambda$$

so we see finally that x lies in $B_0 \cup \cdots \cup B_{\lambda-1}$ as asserted in Formula 6.3.

The existence of the chain complex Formula 6.2 follows in a standard way from this cell decomposition. To write the complex explicitly we need to know the degrees of the 'attaching maps' by which the boundaries of the cells of B_λ are glued to the $(\lambda - 1)$-skeleton $B_0 \cup \cdots \cup B_{\lambda-1}$. The 'Witten complex' gives a recipe for these in terms of the gradient trajectories connecting critical points – a recipe which is implicit in the two cell decompositions, by ascending and descending sets, above. Suppose b_r, b_s are points of adjacent indices $\lambda - 1, \lambda$. Then the spheres Σ_r, Σ_s are of complementary dimensions in W, and so meet in a finite set of points. In other words there are a finite number of trajectories from b_r to b_s, modulo translation. The orientation data (which we are not discussing in detail) allows us to fix signs to give an algebraic intersection number of the spheres, hence an algebraic 'count' of the gradient trajectories, giving a number $n(b_r, b_s)$. This array of numbers is just what is needed to define the matrix of a linear map

$$\partial : C_\lambda \to C_{\lambda-1},$$

and the conclusion of our discussion is the assertion that this recipe makes (C_*, ∂) into a chain complex with homology the ordinary homology of B. Notice by the way that Poincaré duality is built into this theory: replacing f by $-f$ we get another complex computing the homology of f which is precisely the dual of the original one: this is a manifestation of the two cell decompositions by ascending and descending sets. A consequence of this is that it is largely a matter of choice and convention whether one works with homology or cohomology.

It is probably not necessary to labour the analogy we wish to make between the discussion above and the definition of Floer's homology groups. In place of the function f on the compact manifold B we take the Chern–Simons functional on the space of connections $\mathcal{B}$ over a 3-manifold Y. The critical points are the flat connections and the gradient trajectories are the finite energy instantons over the tube. The analysis of the convergence of integral curves sketched above is parallel to our analysis of 'chain convergence' of instantons in the previous Chapter (and the reader who has followed the proofs there will have no difficulty in filling in the details in the more elementary, finite-dimensional, case). The salient differences – beyond the overriding and obvious one that $\mathcal{B}$ is infinite-dimensional – are

- the Chern–Simons functional is circle-valued rather than real-valued,
- the index is only defined modulo 8,
- the space $\mathcal{B}$ is not a manifold due to the presence of reducible connections.

Putting these aside for the moment, however, there are clearly good grounds for thinking of the Floer groups, formally, as some kind of 'homology' groups of the space $\mathcal{B}$. Of course this conception should not be confused with the ordinary homology groups of $\mathcal{B}$, which are perfectly well-defined. Rather one should think of the Floer homology as being a kind of 'middle-dimensional' homology of $\mathcal{B}$, since the negative subspace of the Hessian of the Chern–Simons functional is infinite-dimensional and can be thought of as roughly 'half' the dimension of the whole space (in the same sense as the Hardy space of positive Fourier series has a right to be thought of as being of roughly 'half' the dimension of $L^2(S^1)$). See [3] for a discussion in a similar spirit.

There are two particular points we want to bring out, against this background. The first is a general principle that, given any construction in 'ordinary' algebraic topology, one can seek an analogue in the Floer theory by expressing the construction in terms of gradient trajectories of a Morse function and taking these over to instanton moduli spaces. We shall see examples of this in Chapter 7 when we consider cup products and phenomena arising from reducible connections (which of course have analogues when one takes quotient spaces of finite-dimensional manifolds by a group action which is not free). The second, which we will explain now, is to see why – independent of any of the details of the situation – one might expect a middle-dimensional homology to be the right setting for discussing gluing problems in Yang–Mills theory.

Consider, then, a closed 4-manifold X decomposed into two pieces by a 3-manifold Y. We would like to understand a moduli space $\mathcal{M}$ of instantons over X in terms of the pieces. Let $\mathcal{L}_i$ be the (infinite-dimensional) moduli space of instantons over the manifold-with-boundary X_i, with smooth boundary values. (These paragraphs are intended to be entirely motivational and schematic, so we ignore all technicalities.) There are boundary value maps $\mathcal{L}_i \to \mathcal{B}_Y$, which we assume to be embeddings. Then (assuming no complications with reducible connections etc.) the moduli space $\mathcal{M}$ can be viewed as the intersection $\mathcal{L}_1 \cap \mathcal{L}_2 \subset \mathcal{B}_Y$. Suppose we are in the case when $\mathcal{M}$ is 0-dimensional. Then we have an invariant of X defined by counting (with signs) the points of $\mathcal{M}$, that is to say the algebraic intersection number of $\mathcal{L}_1, \mathcal{L}_2$ in $\mathcal{B}_Y$. Now in a finite-dimensional situation, when we have submanifolds $L_1, L_2 \subset B$, we know that the algebraic intersection factors through the homology groups, so the data we need to record from L_i is just the homology classes $[L_i] \in H_*(B)$. This suggests that one should seek 'homology groups' $\mathcal{H}_*(\mathcal{B}_Y)$ of $\mathcal{B}_Y$, with an intersection pairing, such that the boundary value spaces carry fundamental classes in the groups and the pairing gives the algebraic intersection number. If one has this then the structure of the resulting theory would be to assign a vector space

$$\mathcal{H}(Y) = \mathcal{H}(\mathcal{B}_Y)$$

to Y, with an intersection pairing, such that

- the manifolds-with-boundary X_i define invariants $\psi_i = [\mathcal{L}_i] \subset \mathcal{H}(Y)$,

- the invariant of X given by counting points of $\mathcal{M}$ is the pairing $\langle \psi_1, \psi_2 \rangle$.

As we shall see, the Floer groups do furnish (at least in favourable cases) a theory of this kind, with one notable difference to be discussed below, and we might interpret this as saying that Floer's construction gives a way of making a rigorous homology theory in which the boundary value spaces carry fundamental classes. One can develop this further by asking the following question. Suppose, in the finite-dimensional situation, we have a Morse function f on B and also a q-dimensional submanifold L_1; how do we identify the fundamental class $[L_1] \in H_q(B)$ in the Witten complex description of the homology? The answer is that one computes the intersection numbers of L with ascending manifolds V_r and then $[L_1]$ is represented by the chain

$$\sum (L_1 \cdot V_r) \langle b_r \rangle \in C_q.$$

(In particular this chain is a cycle, and modulo boundaries, is independent of continuous deformations of L_1.) Now suppose we have another submanifold L_2 of complementary dimension $n - q$; how do we see that the algebraic intersection $L_1 \cdot L_2$ goes over to the Poincaré duality built into the Morse complex? We deform L_1 continuously into $L_1^t = \phi_{-t}(L_1)$ and L_2 into $L_2^t = \phi_t(L_2)$ where t is *large*. Then one sees that, if the manifolds are in general position, the intersection points $L_1^t \cap L_2^t$ are localised in small neighbourhoods of the critical points and the contribution from a critical point p_r is $(L_1 \cdot V_r)(L_2 \cdot U_r)$, so that

$$L_1 \cdot L_2 = \sum (L_1 \cdot V_r)(L_2 \cdot U_r),$$

as required. Now take this over to the infinite-dimensional case. There is no real analogue of the gradient flow, since we cannot solve the initial value problem for the instanton equations, but if ρ is a critical point of the Chern–Simons functional, i.e. a flat connection, we can still make sense of the 'ascending' and 'descending' manifolds belonging to ρ. They just correspond to the connections A over Y for which there is an instanton over one of the half-tubes $Y \times [0, \infty), Y \times (-\infty, 0]$ respectively, with boundary value A and asymptotic to ρ at $\pm\infty$. Thus the intersection points of $\mathcal{L}_1$ with the ascending manifold $\mathcal{V}_\rho$ belonging to ρ correspond to instantons over the complete 4-manifold $\hat{X}_1$, obtained by adding a half-tube to the boundary, which are asymptotic to ρ at infinity. So, if we follow the prescription above from the finite-dimensional case, the chain we expect to associate to X_1 in the Floer complex of Y is just

$$\sum_\rho n_{\rho,X_1}\langle\rho\rangle,$$

where n_{ρ,X_1} is the count of instantons over the complete manifold X_1 which are asymptotic to ρ. Likewise the 'gluing argument' which tells us that the number of points (counted with sign) in the moduli space $\mathcal{M}_{\mathcal{X}}$ is

$$\sum_\rho n_{\rho,X_1} n_{\rho,X_2}$$

is a counterpart of the localisation discussion under the positive and negative flows above. In the 4-manifold case we would deform the metric on X by pulling out a long 'neck'. Equivalently, we attach long tubes $Y\times[0,T], Y\times[-T,0]$ to the boundaries of X_1, X_2 to give Riemannian manifolds-with-boundary X_1^T, X_2^T. The boundary value spaces $\mathcal{L}_i^T$ correspond precisely to the deformations $\phi_{-T}(\mathcal{L}_1), \phi_T(\mathcal{L}_2)$ considered above. (The point is that, while the 'flow' ϕ_t is not really well-defined as a map, the image of $\mathcal{L}_i$ is, so long as t has the appropriate sign.)

We will now go back over one point in the discussion above. Instead of a single homology $\mathcal{H}(Y)$ we really prefer to distinguish two (graded) groups $HF(Y), HF(\overline{Y})$ and in place of the bilinear form on $\mathcal{H}(Y)$ we have a pairing between $HF(Y), HF(\overline{Y})$. This reflects the fact that the boundary value spaces $\mathcal{L}_i$ are fundamentally different kinds of objects; for example one should not expect our theory to give a meaning to the *self-intersection* $\mathcal{L}_1\cdot\mathcal{L}_1$. To see this a little more clearly we can ask what special structure an infinite-dimensional manifold should have for there to be a chance of defining a Floer-like theory. For any connection A over Y we have a self-adjoint linear operator Q_A which, as we have explained in Chapter 2, can be thought of as an endomorphism of the tangent space of $\mathcal{B}$ at $[A]$. Thus we have a splitting

$$T\mathcal{B}_{[A]} = T^+\oplus T^-\oplus T^0, \tag{6.4}$$

where T^+, T^- are the spans of the positive and negative eigenspaces of Q_A respectively, and T^0 is the kernel. We expect the kernel to be trivial for generic A but non-trivial on a codimension-1 subset, and this means that we do not quite get a direct sum decomposition of the tangent bundle. However, following the ideas of Segal [39][Chapter 7], we may consider the set of 'positive' subspaces $V\subset T\mathcal{B}_{[A]}$ which are 'commensurable' with T^+ in the sense that the projection $\pi_+ : V\to T^+$ is

Fredholm and the projection $\pi_- : V \to T^-$ is *Hilbert–Schmidt*. Although this definition depends on the splitting Equation 6.4, it is insensitive to the essentially finite-rank changes which occur when the kernel jumps, so this set of positive subspaces forms a bundle Gr^+ over $\mathcal{B}_\mathcal{Y}$. Then we may consider 'positive' submanifolds $\mathcal{L}_+$ of $\mathcal{B}_Y$, whose tangent space lies everywhere in Gr^+, and similarly negative submanifolds $\mathcal{L}_-$, interchanging signs. Of course the boundary value spaces are examples of these. The intersection of a positive and a negative submanifold, if transverse, is a finite-dimensional manifold in $\mathcal{B}_Y$. We should think of the dual Floer groups $HF_*(Y), HF^*(Y) = HF_*(\overline{Y})$ as homology groups whose cycles are the positive and negative submanifolds, respectively, with the pairing between them given by intersection. One can contemplate abstracting this and studying a general infinite-dimensional manifold $\mathcal{B}$ with this kind of 'Segal structure' on its tangent space. One might hope to construct a corresponding Floer-type theory from some kind of homology classes of positive and negative submanifolds. If one has a function F on $\mathcal{B}$ whose covariant second derivative $\nabla\nabla F$ defines the structure, in the manner of the operators Q_A above, then one can try to set up the corresponding Morse–Witten complex, using gradient paths. A large part of the theory could be set up in this generality – roughly speaking those parts having to do with the local properties of the spaces of trajectories – but the crucial issue of the *compactness* of these spaces lies deeper and is harder to axiomatise.

6.2 The straightforward case

We now resume the technical development. We will work throughout this Chapter with the class of oriented 4-manifolds with boundary, as in the TQFT axioms, but we also require that the boundary be a *disjoint union of homology* 3-*spheres*. We also fix orientations on the homology, in the sense of Chapter 2. Let X be such a 4-manifold, with boundary

$$Y = \bigsqcup_{i=0}^{n} Y_i.$$

We define a $(\mathbf{Z}/8)$-graded vector space

$$HF_*(Y) = HF_*(Y_0) \otimes HF_*(Y_2) \otimes \cdots \otimes HF_*(Y_n) \qquad (6.5)$$

and we want to define a vector

$$\Psi_X \in HF_*(Y).$$

(For simplicity we will work throughout this Chapter with rational coefficients, $HF_*(Y_i) = HF_*(Y_i; \mathbf{Q})$.) Of course, in the case when there are two boundary components this agrees with Formula 6.1 when we make the usual identification

$$HF_*(Y_0 \sqcup Y_1) \cong \mathrm{Hom}(HF_*(Y_0), HF_*(Y_1)),$$

using the duality $HF_*(\overline{Y_0}) = HF_*(Y_0)^*$.

The strategy for defining Ψ_X is to go through a chain

$$\psi_X \in C_*(Y_1) \otimes \cdots \otimes C_*(Y_n),$$

where $C_*(Y_i)$ is the Floer chain complex of the homology 3-sphere Y_i, defined using the critical points of the Chern–Simons functional, if this is non-degenerate, and a small generic perturbation otherwise. This chain is defined in the following fashion. Suppose to begin with that we are in the good situation where the flat connections are non-degenerate, and we fix generic metrics on Y_i. Then, in the familiar way, we choose a generic metric with half-infinite tubular ends on X, having the given metrics on the cross-sections of the ends. For any collection of irreducible flat connections ρ_i over Y_i we have an index

$$i(X; \rho_0, \ldots, \rho_n) \in \mathbf{Z}/8,$$

which is the Fredholm index, modulo 8, of the deformation operator of any $SU(2)$ connection with these flat limits. Our chain ψ_X will be a sum

$$\psi_X = \sum \nu_X(\rho_0, \ldots, \rho_n) \rho_0 \otimes \cdots \otimes \rho_n,$$

where $\nu(\rho_0, \ldots, \rho_n)$ vanishes if $i(X; \rho_0, \ldots, \rho_n) \neq 0$. This is the same as saying that ψ_X lies in a particular degree with respect to the grading on $C_*(Y_0) \otimes \cdots \otimes C_*(Y_n)$, and by comparing with the trivial connection one sees that this degree is

$$\deg(\psi_X) = -3(1 - b_1(X) + b_2^+(X)) \mod 8.$$

If $i(X; \rho_0, \ldots, \rho_n) = 0$ there is an adapted $SU(2)$ bundle E over X with these limits which defines a moduli space of dimension 0, and we define $\nu_X(\rho_0, \ldots, \rho_n)$ to be the number of points in M_E, counted with signs and after perhaps making suitable perturbations. One then needs to show that

- ψ_X is a cycle in the tensor product complex,

- if we change the choice of metric, or perturbation of the instanton equation, over X then ψ_X changes by a boundary in the tensor product complex.

(Notice again that, in the case where there are two boundary components and we interpret ψ_X as a linear map, then the first item above is the same as saying that we get a chain map and the second is the same as saying that the map is unique up to chain homotopy.)

Further, one wants to show that if X_1, X_2 are two 4-manifolds of this type with some common boundary components, say

$$\partial X_1 = L_1 \sqcup \bigsqcup_{i=1}^{k} Y_i,$$

$$\partial X_2 = L_2 \sqcup \bigsqcup_{i=1}^{k} \overline{Y}_i,$$

and if X is the manifold obtained from X_1, X_2 by identifying the common boundary components Y_i then there is a gluing relation

$$\Psi_X = 2^{k-1} \langle \Psi_{X_1}, \Psi_{X_2} \rangle \in HF_*(L_1 \sqcup L_2),$$

where the pairing on the right hand side of this formula is obtained from the obvious contraction on the tensor product of Floer groups. The strategy for proving this gluing relation is to choose metrics on X in which the 'necks' joining X_1, X_2 across the Y_i are very long, and then show that for such metrics one has equality at the chain level. The factor 2^{k-1} above arises because of the centre ± 1 of $SU(2)$. When $k = 1$, so there is a single common boundary component, we have a 'gluing parameter' of ± 1 in the choice of the identification we make when gluing instantons on bundles P_1, P_2 with matching limits. However, this gluing parameter can be cancelled by applying the automorphism group ± 1 to P_1, or P_2. When we have k common boundary components we need to specify gluing data in a copy of the group $(\pm 1)^k$ and the cancellation from the automorphisms on either side reduces this to the quotient by the diagonal action of ± 1, so we can construct 2^{k-1} instantons over X from each pair of instantons over X_i with matching limits.

The strategies we have outlined above should be quite familiar since, as we have said, they follow very closely the work we have already done in detail in Chapter 5. One just applies the same compactness and gluing principles to analyse appropriate 0- and 1-dimensional moduli spaces. With this said, we will omit further discussion of the proofs and move on. The point we need to address now is that the real picture is a little more complicated than what we have outlined. On the one hand it falls *short* of the outline because of some difficulties associated with the trivial connection, and with manifolds X with $b^+(X)$ small. On the other hand the true picture goes *beyond* the outline, in that we can define more invariants, using higher-dimensional moduli spaces (and invariants defined by moduli spaces of connections with trivial limits). Both of these aspects enter already into the theory for closed 4-manifolds and we will begin the next Section by reviewing that theory, making use of a device of Morgan and Mrowka – forming connected sum with auxiliary $\overline{\mathbf{CP}}^2$s – which sidesteps a lot of the difficulties associated with the trivial connection. We shall then go on to develop the theory for manifolds-with-boundary.

6.3 Review of invariants for closed 4-manifolds

In this Section we will summarise material which is discussed in more detail in [17] – although we will work in a slightly more general setting than that reference.

Let X be a closed, connected, oriented 4-manifold with a homology orientation (recall that this means an orientation of the line $\Lambda^{\max} H^1(X) \otimes \Lambda^{\max} H^+(X)$). Let $E \to X$ be an admissible bundle with structure group $U(2)$, as in Section 5.6. Recall that we fix some standard connection on $\Lambda^2 E$, then we call a connection on E an instanton if it induces an ASD connection on the associated $SO(3)$ bundle $\mathfrak{g}_E$ and the given connection on $\Lambda^2 E$. If $b^+(X) > 0$ then for generic metrics on X there are no S^1 instantons with non-zero first Chern class, and for generic metrics on X no reducible instantons on any bundle with the same first Chern class as E. There is a universal $SO(3)$ bundle

$$\mathfrak{g}_{\mathbf{E}} \to M_E \times X.$$

The slant product with the class $-\frac{1}{4}p_1(\mathfrak{g}_{\mathbf{E}})$ defines maps

$$\mu : H_i(X;\mathbf{Q}) \to H^{4-i}(M_E;\mathbf{Q}), \quad i = 0,1,2,3. \tag{6.6}$$

We define $A(X)$ to be the free graded-commutative ring generated by the $H_i(X;\mathbf{Q}), i < 4$, with the grading such that elements of $H_i(X)$ have dimension $4 - i$ in $A(X)$. (So $A(X)$ is the tensor product of an exterior algebra on $H_{\mathrm{odd}}(X)$ and a polynomial algebra on $H_2(X) \oplus \mathbf{Q}$.) Then, taking cup products, we get an induced homomorphism

$$\mu : A(X) \to H^*(M_E).$$

The moduli space M_E has an expected dimension

$$\mathrm{ind}(E) = 2(4c_2(E) - c_1(E)^2) - 3(1 - b^1 + b^+(X)). \qquad (6.7)$$

If $4c_2 - c_1^2 \neq 0$ we know that, for generic metrics on X, the moduli space is a smooth manifold of this dimension; and in any case one can choose some auxiliary perturbations of the ASD equations as in Chapter 5 to achieve this condition on a slightly perturbed moduli space, which we will not distinguish in our notation. The homology orientation of X induces an orientation of the moduli space. Now we define invariants formally by evaluating cohomology classes on a fundamental class of the moduli space

$$\langle \lambda \rangle_E = \langle \mu(\lambda), [M_E] \rangle, \quad \lambda \in A(X). \qquad (6.8)$$

This is completely straightforward if the moduli space is compact, which will happen if the dimension is less than 8, but in general it will not be and one needs to specify precisely what the evaluation means. Different approaches to doing this are discussed in [17] (which concentrate on the powers of the 2-dimensional classes – the most interesting for applications). One approach is to show that the classes extend over the natural compactification of M_E. A more elementary approach, which we will stick to here, is as follows. Let $\lambda = \sigma_1 \dots \sigma_k$ be a class in $\Lambda(X)$, so σ_i are homology classes in X. Choose generic cycles representing these classes and small neighbourhoods U_i of these cycles. Irreducible instantons over X remain irreducible over these neighbourhoods so each cohomology class $\mu(\sigma_i)$ can be pulled back to M_E under the map defined by restriction of connections to U_i. Then a rather general argument shows that one can choose generic cochains on the spaces of irreducible connections over the U_i such that the pull-backs give representative cochains whose cup product is a compactly supported representative for $\mu(\sigma_1) \dots \mu(\sigma_k) = \mu(\lambda)$. The pairing Equation 6.8 is then defined by evaluating this compactly- supported cochain on the fundamental class. The scheme sketched here is worked out in detail in [17] for the even-dimensional generators, and when X is simply connected. The use of the odd cohomology classes makes no difference at all: the only minor point which arises in making this extension is that, if one is in the situation where a perturbation of the equations is needed, one needs to arrange the perturbation and the neighbourhoods N_i, so that the solutions of the deformed equation are irreducible.

Now suppose we vary the metric on X (or the perturbations used to define the instanton equations) in a one-parameter family. Then a standard cobordism argument shows that the pairings we have defined do not change, *so long as we do not encounter any reducible solutions in the family.* If $b^+(X) > 1$ then this holds for generic paths, so to sum up for this case we have

Theorem 6.1 *Let X be a closed oriented 4-manifold with a homology orientation, and with $b^+(X) > 1$. If E is an admissible $U(2)$ bundle over X, evaluation on the moduli space M_E defines a map, depending only on the smooth structure of X and the given topological data from $A_d(X)$ to $\mathbf{Q}$, where the degree d is* $\mathrm{ind}(E)/2$.

If $b^+(X) = 1$ the theory is considerably more complicated. One can predict precisely when the reductions will occur, in terms of the periods of the self-dual harmonic form on X. Let $\mathcal{H}_X$ be the quotient of the positive cone

$$\{\omega \in H^2(X;\mathbf{R}) : \omega^2 > 0\}$$

by the action of the scalars (which is naturally a hyperbolic space). A metric g on X determines a point $\omega_g \in \mathcal{H}_X$, via its self-dual harmonic form. For a given bundle E there is a collection of 'walls', making up a subset $W_E \subset \mathcal{H}_X$ such that if ω_g is not in W_E no bundle E' with $c_1(E') = c_1(E)$ and $c_2(E') \leq c_2(E)$ admits a reducible instanton. Let $\mathcal{C}_X$ be the set of connected components of the complement of the walls $\mathcal{H}_X \setminus W_E$: this is a collection of open 'chambers' in the hyperbolic space. Then Kotschick and Morgan [28] have shown that the invariant defined by a generic metric g *depends only on the chamber containing* ω_g. (This is fairly straightforward for low-dimensional moduli spaces, but the general case involves substantial technical difficulties.) To sum up we have

Theorem 6.2 *If X and E are as in Theorem 6.1, except that now $b^+(X) = 1$, then evaluation on the moduli spaces defines a map on the set of chambers*

$$\gamma_X : \mathcal{C}_X \to \mathrm{Hom}(A_d(X), \mathbf{Q}).$$

We will now explain, following Morgan and Mrowka [35], how to use a 'stabilisation' to embed other invariants in the family we have defined above. Let $\overline{\mathbf{CP}}^2$ be the complex projective plane with the reverse of the standard orientation, and let V be the rank-2 bundle over $\overline{\mathbf{CP}}^2$ given by the direct sum $V = \mathbf{C} \oplus L$, where $c_1(L)$ is the Poincaré dual of the standard generator e for $H_2(\mathbf{CP}^2)$. Then if E is *any* $U(2)$ bundle over a 4-manifold X the 'connected sum' bundle $E \sharp V$ over $X \sharp CP^2$ is admissible, even if E itself is not. A homology orientation of X induces one on $X \sharp CP^2$ and the dimension $\mathrm{ind}(E \sharp V)$ is $\mathrm{ind}(E) + 2$. Plainly $A(X \sharp CP^2)$ is obtained by adjoining the element e, of degree 2, to $A(X)$.

Proposition 6.3 *Let X, E be as in Theorem 6.1, so E is admissible, and $b^+(X) = b^+(X \sharp CP^2) > 1$. Then for any $\lambda \in A_d(X)$, where $d = \mathrm{ind}(E_0)/2$, we have*

$$\langle \lambda \rangle_E = \frac{1}{2} \langle e\lambda \rangle_{E \sharp V}.$$

We will recall in outline the proof of Proposition 6.3. Suppose for simplicity that $\mathrm{ind}(E) = 0$, so the invariant in question is just a number obtained by counting points in the discrete set $M_{E_0} = \{[A_\alpha]\}$, say. There is a single instanton connection A_V on the bundle V over CP^2 and this is reducible, induced by the splitting $V = \mathbf{C} \oplus L$. We consider a metric on $X \sharp CP^2$ of the familiar kind, in which the neck of the connected sum is very long. Then the standard 'gluing arguments' show that the connections in the moduli space $M_{E_0 \sharp V}$ are obtained by gluing the A_α to A_V across the connected sum. There is a gluing parameter of $SO(3)$ but this has to be divided by the isotropy group $\Gamma_{A_V} \cong S^1$. So we find in sum that the moduli space $M_{E \sharp V}$, for such a metric, is a union of components S_α, each of which is a copy of $S^2 = SO(3)/S^1$. A calculation shows that

$$\langle \mu(e), S_\alpha \rangle = \pm 2,$$

and one shows also that the signs match up. Summing over α this establishes the desired formula.

Now suppose that E is *not* an admissible bundle over X; for example E might be an $SU(2)$ bundle. Then $E \sharp V$ is admissible over the connected sum and the right hand side of the formula in Proposition 6.3 is still defined. Thus is this case we *define* invariants of X by

$$\langle \lambda \rangle_E = \frac{1}{2} \langle e\lambda \rangle_{E \sharp V} \tag{6.9}$$

So in sum we obtain

Theorem 6.4 *If X is a manifold with $b^+(X) > 1$ and a homology orientation, and if E is any $U(2)$ bundle over X, there is an invariant $\langle\ \rangle_E : A(X) \to \mathbf{Q}$, defined by the moduli space of the connected sum.*

This stabilisation procedure may seem mysterious. The background to have in mind is that if E is not an admissible bundle, for example an $SU(2)$ bundle, the procedure we have used to define invariants still works so long as the second Chern class is large enough: in the 'stable range' discussed in [17]. The same argument as above shows that in this case the two approaches agree. Below the stable range it is plausible that one could still extract invariants by adding on counter terms, in the manner of [14]. However, the Morgan and Mrowka device allows one to avoid all of these complications. It would be possible to combine this discussion with that of chambers for the exceptional case when $b^+(X) = 1$, but we will not go into this here.

A great benefit of the Morgan and Mrowka stabilisation trick is that it allows one to give a very simple proof of the vanishing theorem for connected sums:

Proposition 6.5 *If the manifold X of Theorem 6.4 is a connected sum $X = X_1 \sharp X_2$ and if $b^+(X_1), b^+(X_2)$ are both positive then for any bundle E over X the invariant $\langle\ \rangle_E$ vanishes identically.*

The proof is the usual argument, as in [17], reducing to the case of admissible bundles and showing that for a metric with a long neck the support of a representative for $\mu(\lambda)$ in the moduli space is actually empty. The point is that one can arrange that no reducible connections appear, which made for the complications in the earlier proofs. As we have mentioned at the beginning of this Chapter, this vanishing theorem makes one starting point for the discussion of Floer homology. We can think of Floer's theory as an answer to the question of how Proposition 6.5 changes if in place of ordinary connected sums we allow generalised splittings across other 3-manifolds.

6.4 Invariants for manifolds with boundary and $b^+ > 1$

Now let X be a (connected) 4-manifold as considered in Section 6.2, with boundary a disjoint union of homology spheres, with a homology orientation and with $b^+(X) > 1$. Let v be any class in $H^2(X; \mathbf{Z})$. This class determines, up to isomorphism, a topological $U(2)$ bundle E over X (which of course reduces to $SU(2)$ over each boundary component).

The main results of this Section are as follows.

Theorem 6.6 *The instanton moduli spaces over X define a map*

$$\Psi_X = \Psi_{v,X} : A(X) \to HF(\partial X),$$

such that

$$\deg(\Psi(\lambda)) = \deg(\lambda) - 3(1 - b^1(X) + b^+(X)) \mod 8,$$

which is an invariant of the smooth manifold X, the class v and the homology orientation.

For tidiness, we can also include here the case when the boundary is empty, that is, when X is a closed 4-manifold, if we define the Floer homology of the empty set to be $\mathbf{Q}$ and use the invariants of the previous Section. (In this case there are different bundles E with the given first Chern class, but the grading takes care of this.)

Let X_1 and X_2 be two such connected 4-manifolds with k diffeomorphic boundary components Y_i as in Section 6.2, and define $X = X_1 \cup_{Y_i} X_2$ by identifying these boundary components. The inclusion maps $H_i(X_1) \oplus H_i(X_2) \to H_i(X)$, $i = 0, 1, 2, 3$, define an obvious map

$$\sigma : A(X_1) \otimes A(X_2) \to A(X),$$

and we also have the contraction map on Floer homology

$$\tau : HF_*(\partial X_1) \otimes HF_*(\partial X_2) \to HF_*(\partial X).$$

Theorem 6.7 *Let v_1, v_2 be classes in $H^2(X_1), H^2(X_2)$ and $v = v_1 + v_2 \in H^2(X)$. Then*

$$\Psi_{v,X} \circ \sigma = 2^{k-1} \tau \circ (\Psi_{v_1,X_1} \otimes \Psi_{v_2,X_2}).$$

Of course, one should not let the algebraic formulation disguise the real essence of the assertion, which is that one can compute pairings for moduli spaces over X by decomposing the homology classes into components in X_1, X_2 and then using relative invariants for these two pieces, with values in the Floer homology. Notice that a particular case of Theorem 6.7, when the boundary of X is empty, gives a relation between the invariants of Section 6.3, for closed manifolds, and the new invariants we are considering in this Section.

Having set up these main results – one of the main goals of this book – we hope that there is little need to say anything about the proofs, which are straightforward combinations of the proofs and definitions for closed manifolds outlined in the previous Section, and the basic Floer homology constructions we have seen in Chapter 5.

7
Reducible connections and cup products

We have now seen how the Floer groups fit into a topological field theory for a restricted class of 4-manifolds: with $b^+ > 1$ and with homology sphere boundaries. The difficulties in extending this theory to cover a wider range arise from the presence of reducible connections, either as instanton solutions over 4-manifolds or as flat connections over their boundaries, and one theme of this Chapter is to explore refinements of the theory which take these into account. The other theme is to develop the product structure on the Floer groups, from various points of view. These products interact with the reducible connections in an interesting way, and we shall develop some algebro-topological machinery to handle this. As usual we shall emphasise the case of homology 3-spheres, although making some remarks about more general 3-manifolds. At the end of the Chapter we will bring the ideas together to discuss the Floer homology of connected sums. Throughout this Chapter the symbol Y will denote an oriented homology 3-sphere, and we use Floer groups with rational co-efficients.

7.1 The maps D_1, D_2

In Chapter 5 we defined the Floer boundary map $\partial : C_i \to C_{i-1}$ by counting instantons over the tube with irreducible limits. We ignored the reducible connection θ. We now bring it into the picture by defining maps

$$D_1 : C_1(Y) \to \mathbf{Q}, \quad D_2 : \mathbf{Q} \to C_{-4}(Y).$$

These are defined in just the same way as the boundary ∂. If $\langle \rho \rangle$ is a generator of $C_1(Y)$ there is a 1-dimensional moduli space $M_{\rho\theta}$ of instantons over the tube with limit ρ at $-\infty$ and θ at $+\infty$: this follows from the way we fixed the grading of the Floer chains. We count, with suitable signs, the points in the translation-reduced space $M'_{\rho\theta}$ to obtain a number n_ρ and define $D_1(\langle\rho\rangle) = n_\rho$. Similarly for the map D_2. The same argument that established that $\partial^2 = 0$ gives the relations

$$D_1\partial = \partial D_2 = 0.$$

We define a $(\mathbf{Z}/8)$-graded chain complex $\overline{CF}_*(Y)$ with chain groups

$$\overline{CF}_i = C_i, \; i \neq 0, \quad \overline{CF}_0(Y) = C_0 \oplus \mathbf{Q}, \tag{7.1}$$

and differential $\partial + D_1$. This gives homology groups $\overline{HF}_*(Y)$ (which are independent of the various choices made, as the reader may wish to check now). The ordinary Floer complex is a quotient of $\overline{C}_*$, and we have an exact homology sequence

$$0 \to \overline{HF}_0 \to HF_0 \to \mathbf{Q} \to \overline{HF}_{-1} \to HF_{-1} \to 0,$$

while $\overline{HF}_i = HF_i$ for $i \neq 0, -1$.

We can use the map D_2 in a similar way to define a complex with groups

$$\underline{CF}_i(Y) = \underline{CF}_i, \; i \neq -3, \quad \underline{CF}_{-3}(Y) = C_{-3} \oplus \mathbf{Q} \tag{7.2}$$

and differential $\partial + D_2$. This gives homology groups $\underline{HF}_*(Y)$, and there is an induced map $HF_*(Y) \to \underline{HF}_*(Y)$. Now recall that if, as usual, $\overline{Y}$ is Y with the opposite orientation, we have a dual pairing:

$$CF_i(Y) \otimes CF_{-3-i}(\overline{Y}) \to \mathbf{Q}.$$

Changing orientations, and taking the adjoints, interchanges D_1 and D_2, so we get a perfect pairing

$$\overline{HF}_*(Y) \otimes \underline{HF}_*(\overline{Y}) \to \mathbf{Q}, \tag{7.3}$$

which is compatible in an obvious way with the pairing on the ordinary Floer groups and the maps $\overline{HF}_*(Y) \to HF_*(Y), HF_*(\overline{Y}) \to \underline{HF}_*(\overline{Y})$.

7.2 Manifolds with $b^+ = 0, 1$

Now suppose that Y is the boundary of an oriented 4-manifold X_1 with $b^+(X_1) > 1$. We define a map $\psi_0 : A(X_1) \to \mathbf{Q}$ in the familiar fashion, evaluating classes on moduli spaces of connections with boundary value the trivial connection θ over Y. Combining this with the previous map $\psi : A(X_1) \to C_*(Y)$ we get a map $\overline{\psi} : A(X_1) \to$

$\overline{CF}_*(Y)$. Our gradings have been defined so that $\overline{\psi}$ shifts gradings by $3(1 - b_1(X_1) + b_+(X_1))$ mod 8.

Proposition 7.1 *For $\lambda \in A(X_1)$, the class $\overline{\psi}(\lambda)$ is a cycle for $\partial + D_1$, and the classes defined by different metrics on X_1 differ by a $(\partial + D_1)$-boundary.*

Thus we conclude that, passing to homology, we have a map

$$\overline{\Psi} : A(X_1) \to \overline{HF}_*(Y), \tag{7.4}$$

which plainly lifts our former invariant Ψ over the natural map from $\overline{HF}_*(Y)$ to $HF_*(Y)$. The proof of Proposition 7.1 is straightforward. The point is that when we consider θ as a boundary value it behaves no differently from the irreducible flat connections, so the reader can copy the proof of Theorem 6.4 word for word. To see that $\Psi(\lambda)$ is a cycle we consider moduli spaces of dimension $\deg \lambda + 1$ and boundary value θ, cut down by a representative for λ of codimension $\deg \lambda$. This gives a 1-dimensional space with ends which correspond to factorisations into a connection over X_1 with boundary value ρ, for $\langle \rho \rangle \in C_1(Y)$, and an instanton on the tube running from ρ to θ. The assertion follows by counting the ends, since these instantons on the tube are precisely the data used to define D_1. Similarly, the variation under the change of metrics follows by considering the ends in one-parameter families of moduli spaces with boundary value θ and dimension $\deg \lambda$.

Now if $X = X_1 \cup_Y X_2$ is a 4-manifold split into two pieces, in the familiar way, and if $b^+(X_1), b^+(X_2)$ are both greater than 1, we can if we want cast the basic gluing relation in the following way. Composing the pairing Formula 7.3 with the natural maps

$$\overline{HF}_*(\overline{Y}) \to HF_*(\overline{Y}) \to \underline{HF}_*(\overline{Y}),$$

we get a pairing

$$\overline{HF}_*(Y) \otimes \overline{HF}_*(\overline{Y}) \to \mathbf{Q}.$$

We have invariants

$$\overline{\Psi}_{X_1} : A(X_1) \to \overline{HF}_*(Y), \quad \overline{\Psi}_{X_2} : A(X_2) \to \overline{HF}_*(\overline{Y}),$$

and we have a gluing relation

$$\Psi_X = \langle \overline{\psi}_{X_1}, \overline{\psi}_{X_2} \rangle. \tag{7.5}$$

At this stage, however, we have not really gained anything by this extra complication. The point is that, since the pairing on $\overline{HF}_*(Y) \otimes \overline{HF}_*(\overline{Y})$

factors through the familiar pairing on $HF_*(Y) \otimes HF_*(\overline{Y})$, the gluing formula Equation 7.5 is a consequence of the earlier result Theorem 6.7, and hence contains no more information. The invariants $\overline{\Psi}_{X_i}$ *do*, potentially, contain more information than the Ψ_{X_i} – they record data from the moduli spaces of connections with trivial boundary values – but this is thrown away in our gluing formula. This is just the same mechanism that gives the vanishing theorem for connected sums. We will now go on to see how this formalism allows us to extend the theory to manifolds with $b^+ = 0, 1$, where the trivial connection plays an active role.

7.2.1 The case $b^+ = 1$

Consider a closed manifold $X = X_1 \cup_Y X_2$ where $b^+(X_1) = 1$ and $b^+(X_2) > 1$. We have to reconcile the conflict between the facts that

- since $b^+(X_1) = 1$ we encounter difficulties in defining invariants for X_1, because of reducible connections appearing in one-parameter families of metrics, as we have described in Section 7.1 in the case of closed 4-manifolds,
- we *do* have straightforward invariants for the closed manifold X, since $b^+(X) > 1$, and we expect a gluing relation expressing these in terms of data over X_1, X_2.

This apparent conflict is resolved using the formalism above. Consider the manifold X_1, with $b^+(X_1) = 1$ and $\partial X_1 = Y$. For generic metrics g on X_1 there are no non-trivial reducible connections so we can carry through the theory just as before to define a map

$$\Psi_{X_1,g} : A(X_1) \to HF_*(Y),$$

which will only change in a one-parameter family g_t at parameter values t for which there are integral anti-self-dual harmonic forms on X_1. For each, generic, metric the discussion of the gluing relation goes through just as before, so we have

$$\Psi_X = \langle \Psi_{X_1,g}, \Psi_{X_2} \rangle. \tag{7.6}$$

Now recall that we have a map from $HF_*(Y)$ to $\underline{HF}_*(Y)$. We define $\underline{\Psi}_{X_1,g}$ to be the composite of $\Psi_{X_1,g}$ with this map. The main result of this Section is

Proposition 7.2 *For $\lambda \in A(X_1)$ with $\deg \lambda < 8$ the class $\underline{\Psi}_{X_1,g}(\lambda)$ is independent of the generic metric g on X_1.*

Given this, we obtain a metric-independent invariant $\underline{\Psi}_{X_1}$ – at least in the range $\deg \lambda < 8$ – and in this range we have a satisfactory gluing formula

$$\Psi_X = \langle \underline{\Psi}_{X_1}, \overline{\Psi}_{X_2} \rangle,$$

which follows from Equation 7.6, and the compatibility between the different Floer groups.

The content of Proposition 7.2 is that the chain $\psi(\lambda)$ changes by classes in the image of the D_2 map for Y. The restriction $\deg \lambda < 8$ in Proposition 7.2 is almost certainly unnecessary. We include it in order to avoid complications in our analysis below of the ends of moduli spaces – complications which are very similar to those we referred to in Section 7.1 in the theory for closed manifolds. Very likely the techniques used by Kotschick and Morgan in [28] can be used to remove this restriction, but we will not go into this here.

We will just give the proof of Proposition 7.2 in the case of 0-dimensional moduli spaces: i.e. we consider the image of the identity in $A(X_1)$ under $\Psi_{X_1,g}$. We will also suppose that X_1 is simply connected. The general case is similar. Let g_t, $0 \leq t \leq 1$, be a generic one-parameter family of metrics and ρ a flat connection over Y of Floer degree $3(1 - b_1(X) + b_+(X)$ – so there is an adapted bundle over X_1 with boundary value ρ and index 0. For generic parameter values t the corresponding moduli space $M(t)$ is a finite set, and defines a number $n_\rho(t)$, counting the points with signs. There are two reasons why the numbers $n_\rho(t)$ can change. First, we may encounter irreducible connections in moduli spaces of virtual dimension -1, just as in the ordinary case, and this allows the chain $\sum n_\rho(t)$ to change by the image of ∂ in the familiar way. What we want to examine now is the new mechanism by which $n_\rho(t)$ can change, due to the reducible solutions, so we may as well suppose that in the interval $[0, 1]$, there are no irreducible solutions of index -1 but the period point of $\omega(g_t)$ crosses a single wall defined by a class $e \in H^2(X; \mathbf{Z})$, at time $t = \frac{1}{2}$ say. Thus there is a reducible solution A_0 over X_1, with the metric $g_{\frac{1}{2}}$, having the trivial connection θ as boundary value. Examining the usual compactness argument we see that any change in $n_\rho(t)$ across $t = \frac{1}{2}$ arises due to factorisations in which we have a non-flat reducible instanton A_0 over X_1 with limit θ glued to an instanton I on the tube running from θ to ρ, where

$$\mathrm{ind}(A_0) + \mathrm{ind}(I) + 3 = 0.$$

(In the usual way, we may contemplate more complicated factorisations but these will be ruled out in the end by index considerations.) Now, since A_0 is not flat, ind A_0 is at least

$$2 - 3(1 - b_1 + b_+) = -4,$$

so we see that the only possibility is that $\operatorname{ind}(A_0) = -4$, $\operatorname{ind}(I) = 1$. Thus the instanton I on the tube occurring in the factorisation is one of those which are counted in defining the map D_1.

We now pass on to the reverse step. Suppose A_0 is a reducible instanton over X_1, for the metric $g_{\frac{1}{2}}$, and I is an instanton over the tube with index 1 running from θ to ρ. We consider the problem of gluing these two connections together to make a solution of the instanton equations with respect to g_t, where t is near $\frac{1}{2}$. The deformation complex for A_0 has the form

$$H^1_{A_0} = \mathbf{C}^p, H^2_{A_0} = \mathbf{R} \oplus \mathbf{C}^{p+1},$$

where the stabilser $\Gamma_{A_0} = S^1$ acts in the standard way on $\mathbf{C}$ and trivially on $\mathbf{R}$. We may suppose that $p = 0$, by choosing our path of metrics to be sufficiently generic. Then to analyse the gluing problem we introduce a space $V = S^2 \times (L, \infty)$, where $S^2 = SO(3)/S^1$ is the effective parameter in gluing the two connections across the trivial connection θ over Y, and the factor (L, ∞) contains the usual parameter T measuring the separation in the sum. The component $\mathbf{C}$ in $H^2_{A_0}$ defines a bundle $\mathcal{H} \to \mathcal{V}$, in fact just the pull-back of the tangent bundle over S^2, and our general gluing machinery says that there is a family of sections σ_t of $\mathcal{H}$ and maps $f_t : V \to \mathbf{R}$ such that for each time t near $\frac{1}{2}$ the common zeros of σ_t and f_t model the g_t instantons which can be formed by gluing A_0 to I.

The essential point now is that $f_t(x, T)$ has a limit a_t as $T \to \infty$, and this limit is just the function associated to the deformation problem over X_1. In particular one finds that, if the path g_t is transverse to the wall associated to A_0, then a_t vanishes at $t = \frac{1}{2}$, but $a'(\frac{1}{2}) \neq 0$. It follows that for any fixed large enough T we can find $\delta > 0$ such that for all $x \in S^2$, the function $f_t(x, T)$ changes sign, with non-zero derivative, in the interval $(\frac{1}{2} - \delta, \frac{1}{2} + \delta)$. So for fixed x and T there is some (unique) parameter $t = t(x, T)$ near $\frac{1}{2}$ such that $f_t(x, T) = 0$. As $T \to \infty$ the function $t(x, T)$ tends to $\frac{1}{2}$ from either above or below. Let us suppose that $t(x, T) < \frac{1}{2}$. For fixed T the other component σ_t represents a vector field on S^2, and so the algebraic sum of its zeros represents the Euler number, +2, of S^2. We conclude then that, counted with signs, there are two families of common zeros $(x(t), T(t))$ of f_t, σ_t for t slightly less than $\frac{1}{2}$, which diverge as $t \to \frac{1}{2}$ in the sense that $T(t) \to \infty$ as $t \to \frac{1}{2}$. Hence we conclude that the change in the number $n_\rho(t)$ associated to the splitting of this kind is ± 2. Taking account of signs and summing over all the instantons I of index 1 we find that $\psi(g_t)$ changes by $2D_2(1) \in C_{-4}$.

7.2.2 The case $b^+ = 0$

We will now discuss how the theory of Chapter 6 can be modified for manifolds with $b^+ = 0$. This discussion will be taken further in Subsection 7.3.5 below. To avoid complications we will work with simply connected 4-manifolds, as in the previous Subsection, although again this is not really essential.

Begin by considering the case when X_1 is a simply connected 4-manifold with one boundary component Y, and $b^+(X_1) = 0$. Fix a first Chern class $v_1 \in H^2(X_1)$; then for each irreducible flat connection ρ over Y we can define a number n_ρ by counting instantons in 0-dimensional moduli spaces, as usual, and thus a Floer chain ψ. The compactness argument which proves that, modulo $\operatorname{Im}\partial$, the class ψ is independent of metric goes through. The only new phenomenon we have to consider here are splittings in which an instanton over X_1 with limit θ is glued to an instanton I over the tube from θ to ρ, where A might be reducible. But then we have

$$\operatorname{ind} A + 3 + \operatorname{ind} I = 0$$

and this possibility is ruled out since $\operatorname{ind} A \geq -3$ for any instanton (with equality if and only if A is trivial), while $\operatorname{ind} I \geq 1$. The usual theory must, however, be modified in two ways.

- If ψ is in CF_{-3} then $\partial\psi$ need not vanish. For in the usual proof one looks at the ends of 1-dimensional moduli spaces $M(\sigma)$, for $\langle\sigma\rangle \in CF_{-4}$. However, if the Chern class v_1 is zero (i.e. we are considering $SU(2)$ bundles) there can now be other ends of this space resulting from splittings where the trivial connection over X_1, with index -3, is glued to an instanton I over the tube with index 1 running from θ to σ. These instantons over the tube are just the ones which are used to define the map D_2, and we conclude that, when P is trivial,

$$\partial\psi = -D_2(1) \in CF_{-4}. \tag{7.7}$$

Thus the chain $\underline{\psi} = \psi \oplus (1) \in CF_* \oplus \mathbf{Z} \cong \underline{CF}_*$ is a cycle, and the preceding discussion tells us that the homology class of ψ is metric-independent.

- If $v_1^2 = 2 \mod 4$ and ψ is in CF_1 then $D_1\psi$ need not vanish. Again, the usual proof of this fact involves looking at the ends of a 1-dimensional moduli space $M(\theta)$, with θ as boundary value. But now M_θ may contain points corresponding to reducible connections over X_1, and these will lead to extra boundary components, just as one studies for closed 4-manifolds with $b^+ = 0$ [19]. We conclude that

$$D_1\psi = S(X_1, v_1), \tag{7.8}$$

where $S(X_1, v_1)$ counts, with signs, the relevant reductions. These reductions are labelled by classes $e \in H^2(X_1; \mathbf{Z})$ with $e^2 = -2$ and $e = v_1 \mod 2$, where e and $-e$ define the same reduction. The sign, $\sigma(e)$, associated to the reduction is $+1$ if $(\frac{e-v_1}{2})^2$ is even and -1 otherwise. So the contribution is

$$S(X_1, v_1)) = \tfrac{1}{2} \sum_e \sigma(e), \tag{7.9}$$

where the sum runs over classes e with $e^2 = -2$, equal to v_1 modulo 2.

So to sum up we have

Proposition 7.3 *If X_1 is simply connected with boundary the homology sphere Y, and $b^+(X_1) = 0$, then for a class $v_1 \in H^2(X_1)$ we have Floer classes*

- $\Psi_{X_1,v_1} \in HF_*(Y)$, *if $v_1 \neq 0 \mod 2$, and if $v_1^2 = 2 \mod 4$ this satisfies*

$$D_1(\Psi_{X_1,v_1}) = S(X_1, v_1),$$

- $\underline{\Psi}_{X_1,v_1} \in \underline{HF}_*$, *if $v_1 = 0 \mod 2$.*

The first case above gives a simple generalisation of results on the non-existence of *closed* 4-manifolds with non-standard intersection forms:

Corollary 7.4 *If $S(X_1, v_1) \neq 0$ then the Floer group $HF_1(Y) \neq 0$.*

The point here is that the number $S(X_1, v_1)$ depends only on the intersection form of X_1, and the chosen class v_1. For example, suppose the intersection form of X_1 is the standard 'diagonal' form. It is easy to see that the only case when reductions occur is when the form has rank 2 and $v_1 = (1,1)$ in the standard basis. Then there are two reductions, labelled by

$$\pm(1,1), \pm(1,-1),$$

and the sign σ is $+1$ on the first term and -1 on the second, so we see that $S(X_1, v_1)$ is always zero if the intersection form is standard, and indeed this must be so by Corollary 7.4 since in that case we can take X_1 to be a connected sum of copies of $\overline{\mathbf{CP}}^2$ minus a ball, in which case $Y = S^3$ and $HF_*(Y) = 0$. On the other hand if we take the non-standard form E_8, for example, one readily checks that, for a suitable v_1, we have $S(X_1, v_1) = 1$, so we conclude that if Y bounds a 4-manifold with this intersection form then $HF_1(Y) \neq 0$.

We now move on to the gluing rules in this context. Suppose X_1 is as above and occurs in a decomposition $X = X_1 \cup_Y X_2$ of a closed, simply connected 4-manifold. If $b^+(X_2)$ is also zero, then $b^+(X) = 0$ and we have not defined any invariants for X, so nothing more needs to be said. If $b^+(X_2) > 1$ we have

Proposition 7.5 *If* $b^+(X) = b^+(X_2) > 1$ *then the invariant* $\Psi(X) \in \mathbf{Q}$, *defined by* $v_1 + v_2 \in H^2(X)$, *and* 0*-dimensional moduli spaces, is given by*

$$\Psi(X) = \langle \Psi(X_1), \Psi(X_2) \rangle$$

or

$$\Psi(X) = \langle \underline{\Psi}(X_1), \overline{\Psi}(X_2) \rangle$$

in the two cases of Proposition 7.3.

We will leave the interested reader to formulate a similar statement for the case when $b^+(X_2) = 1$ and the invariants depending on chambers.

The proofs here do not differ in substance from those of the usual gluing formulae. The point is to see that the pairing between $\underline{HF}_*(Y)$ and $\overline{HF}_*(\overline{Y})$ handles the reductions tidily.

7.3 The cup product

7.3.1 Algebro-topological interpretation

Suppose we have a cobordism W between homology 3-spheres Y_0, Y_1, and that $b^+(W) > 1$. By Theorem 6.6 we have a Floer class $\Psi(\frac{1}{2}\text{Point})$, where Point is the generator of $H_0(W)$. (The factor of $1/2$ is included here to simplify the formulae later: *note that this factor has to be taken into account when comparing our formulae with others in the literature*). In the familiar way this can be viewed as a map on the Floer homology, say $U_W : HF_*(Y_0) \to HF_*(Y_1)$. In this Subsection we extend the discussion to the case when W is the trivial product cobordism. This does not of course fit the hypotheses above due to difficulties from the reducible connection, and the whole point of the work is to examine these in more detail. To begin with however we move to the framework of $U(2)$ connections, so instead of the homology sphere Y we consider a 3-manifold Z with non-trivial H^1, and we work with an admissible bundle over Z, with $c_1 \neq 0$. Then in this case there are no difficulties with reducible connections so the product cobordism gives maps

$$U : HF_i(Z) \to HF_{i-4}(Z).$$

The first thing we want to explain is the formal interpretation of these in terms of the space of connections $\mathcal{B}_Z$. So let us go back to our finite-dimensional analogue of Section 6.1: the compact manifold B with a Morse function f. Suppose we have a cohomology class $[\alpha] \in H^\nu(B)$, represented by some explicit cochain α; for example a closed differential form on B. From algebraic topology one knows that this defines product maps

$$\alpha : H_i(B) \to H_{i-\nu}(B),$$

and we now ask how these maps can be represented in terms of the Morse complex description of $H_*(B)$. (The maps we are considering are 'cap products' on homology: it is perhaps more usual to work with the 'cup products' on cohomology but this merely involves a change of notation, using the Poincaré duality built into the Morse complex.) Let p_r, p_s be critical points of index $i, i-\nu$ respectively and consider the set $P_{r,s} \subset B$ of points which flow as $t \to \infty$ to p_r and as $t \to -\infty$ to p_s: i.e. the intersection of the ascending manifold from p_s and the descending manifold from p_r. If f is suitably generic this is a manifold of dimension ν and its closure $\overline{P_{r,s}}$ in B is given by adjoining the points p_r, p_s and certain other manifolds $P_{r',s'}$ corresponding to 'broken trajectories'.

We evaluate α on $\overline{P_{r,s}}$ to get a number $A_{r,s}$, and we interpret the resulting array of numbers as the matrix of a linear map $A : C_* \to C_{*-\nu}$. We claim then that A is a chain map which induces the cap product on homology. The proof of the first assertion, i.e. that $\partial A = \pm A\partial$ follows the familiar pattern. Given a critical point r_v of index $i - \nu - 1$ we consider the $(\nu + 1)$-dimensional manifold $P_{r,v}$. The boundary of this, as a set, is made up of various pieces, involving the intermediate critical points, but the *codimension*-1 components of the boundary just involve critical points p_u of index either $i-1$ or $i-\nu$:

$$\overline{P_{r,v}} = P_{r,v} \cup \bigcup_{u:i(u)=i-1} P_{u,v} \cup \bigcup_{u:i(u)=i-\nu} P_{r,u} \cup R,$$

where R is contained in a union of manifolds of dimension at most $\nu-2$. From the point of view of homology each codimension-1 component of the boundary must be counted with a multiplicity equal to a (signed) number of flow lines (we are ignoring the issue of orientation throughout this discussion). For example if $i(u) = i-1$ the intersection of the neighbourhood of a point in $P_{u,v}$ with $P_{r,v}$ is a union of a finite number of manifolds-with-boundary, one for each flow line from p_r to p_u, coming together along the boundary. So the boundary in homology of $\overline{P_{r,v}}$ is

$$\partial_{\text{homology}} \overline{P_{r,v}} = \sum_{r:i(r)=i-1} n_{r,u}[P_{u,v}] + \sum_{r:i(r)=i-\nu} n_{u,v}[P_{r,u}]. \qquad (7.10)$$

Now the fact that

$$0 = \langle d\alpha, \overline{P_{r,v}} \rangle = \langle \alpha, \partial_{\text{homology}} \overline{P_{r,v}} \rangle = \sum n_{p,r} A_{r,s} + \sum A_{p,r} n_{r,s}$$

translates into the formula $\partial A \pm A\partial = 0$.

Going back to the space of connections on the 3-manifold Z, we consider the cohomology class $u \in H^4(\mathcal{B}^*_Z)$ defined by $\frac{1}{2}\mu(\text{Point})$. We choose a representative for this defined by restriction to a small neighbourhood N of a point in Z (so strictly this representative will only be defined over the subset of connections which are irreducible over N) and follow though the recipe above, interpreting flow lines as instanton moduli spaces. Then we see that we arrive at exactly the same conclusion as we reached in the 4-dimensional theory of the previous Section – evaluating the cochain on the subset of $\mathcal{B}^*_Z$ is just the same as evaluating on the moduli space via restriction of connections over the tube $Z \times \mathbf{R}$ to $N \times \{0\}$. The upshot is that we interpret the map $U : HF_* \to HF_{*-4}$ as the cap product between the ordinary 4-dimensional cohomology class and the Floer homology, thought of as the semi-infinite-dimensional homology of $\mathcal{B}_Z$.

While this 4-dimensional product will be of primary interest to us, notice that the discussion applies equally well to other homology classes. For any class $z \in H_*(Z)$ we can consider the corresponding class $i_*(z) \in H_*(Z \times [0,1])$ and, applying the theory of Chapter 6, get a map from $HF_*(Z)$ to $HF_{*+i-4}(Z)$ which we interpret as the cap product with $\mu(z)$

in $H^{4-i}(\mathcal{B}^*_{\mathcal{Z}})$. In fact, provided one works with low-dimensional moduli spaces, so that we do not encounter bubbling of instantons, one can mimic the finite-dimensional theory directly; so *any* cohomology class $\alpha \in H^{\nu}(\mathcal{B}^*_{\mathcal{Z}})$ with $\nu < 7$ defines a product map on the Floer groups. (In rational cohomology all classes are generated by the $\mu(z)$, but there could be other torsion classes.)

7.3.2 An alternative description

There is another, more 3-dimensional, point of view which gives a slightly different representation of this cup product at the chain level. Suppose first that we have a compact, oriented, 3-dimensional manifold $\mathcal{M}$ parametrising a family of connections over the tube $Z \times \mathbf{R}$, all with fixed flat limits ρ, σ and normalised to have 'centre of mass' 0, in the sense that

$$\int_{Z\times\mathbf{R}} t|F|^2 = 0. \tag{7.11}$$

Let $\mathrm{ad}\,\rho$, $\mathrm{ad}\,\sigma$ be fixed flat $SO(3)$ bundles over Z corresponding to ρ and σ. Each point $\mathbf{A} \in \mathcal{M}$ gives a one-parameter family A_t in $\mathcal{B}_Z$, as discussed in Chapter 2. Then we can define a map $i : \mathcal{M} \times (-\infty, \infty)$ to $\mathcal{B}_Z$ by

$$i(\mathbf{A}, s) = A_{\tan^{-1}(\pi s/2)}.$$

The assumption on the limits means that this map extends to the suspension

$$\Sigma\mathcal{M} = \mathcal{M} \times [-1, 1]/\sim,$$

where $\sim$ is the usual equivalence relation identifying $\mathcal{M} \times \{\pm 1\}$ to a pair of points. We assume that all the connections A_t are irreducible, so we get a map

$$i : \Sigma\mathcal{M} \to \mathcal{B}^*_Z.$$

We may then pull back the 4-dimensional cohomology class $u \in H^4(\mathcal{B}^*_Z)$ and evaluate this on the fundamental class in $H_4(\Sigma\mathcal{M}) = \mathbf{Z}$. On the other hand, we may fix a base point z_0 in Z and fix frames for the fibres of the $SO(3)$ bundles $\mathrm{ad}\,\rho, \mathrm{ad}\,\sigma$ over the base point. Then the *holonomy* of the connection $\mathbf{A}$ along the line $\{y_0\} \times \mathbf{R}$ gives, using these frames, an element of $SO(3)$, so we get a map

$$h : \mathcal{M} \to SO(3),$$

which has an integer *degree*.

Lemma 7.6 *The degree of the holonomy map h is $4\langle i^*(u), \Sigma\mathcal{M}\rangle$.*

The proof is straightforward when one unwinds the definitions. Pulling back the base point fibration over $\mathcal{B}_Z^*$ gives an $SO(3)$ bundle over $\Sigma\mathcal{M}$ and the frames we have chosen are frames for this bundle over the two vertices of this suspension. These extend to trivialisations of the bundle over the two cones making up the suspension, and the transition function is precisely h. Thus the assertion is the standard 'transgression' formula, expressing the characteristic class of a bundle over a suspension in terms of the homotopy class of the transition function.

In our application we want to consider the case where $\mathcal{M}$ is a (reduced) moduli space of instantons over the tube. The number $\langle i^*(u), \Sigma\mathcal{M}\rangle$ is the same as the number we obtain by evaluating the 4-dimensional class in defining the chain map U above. So, when all these moduli spaces are compact, we can immediately use Lemma 7.6 to give another description of U. The complication comes from the fact that these moduli spaces will not normally be compact and we have to modify the holonomy construction to get a genuine chain map. For simplicity we suppose that we can lift our frames for the bundles over the base point to $U(2)$ (the extension of the construction to avoid this assumption is left as an exercise for the reader). We choose such frames for all flat connections over Y. Since the determinants of our connections are fixed we get a holonomy map into a copy of $SU(2)$ inside $U(2)$. So for any reduced moduli space $M'_{\rho\sigma}$ we have a map

$$\tilde{h} : M'_{\rho\sigma} \to SU(2).$$

Our first thought is to choose a representative α for the generator of $H^3(SU(2))$, pull this back and evaluate on the 3-dimensional moduli spaces to obtain the matrix entries of a linear map L on the Floer chains in the usual way; but unfortunately this need not give a chain map. To see why consider a pair of flat connections ρ, τ of index difference 5. Thus there is a 4-dimensional (reduced) moduli space $\mathcal{M}^4$ of instantons from ρ to τ. The compactification is a manifold-with-corners and the codimension-1 faces are of four kinds, corresponding to factorisations through a connection σ with $\operatorname{ind}\sigma = \operatorname{ind}\rho - i$ for $i = 1, 2, 3, 4$. The holonomy map extends to the compactification and

$$\tilde{h}_{\rho\tau} = \tilde{h}_{\rho\sigma}\tilde{h}_{\sigma\tau}, \tag{7.12}$$

on the face $M'_{\rho\sigma} \times M'_{\sigma\tau} \subset \overline{M'_{\rho\tau}}$.

We can pull α back to all of these faces and evaluate, and we know that the sum of all these contributions is zero. The sum of contributions from factorisations with $i = 1$ and $i = 4$ gives the matrix entries of $\partial L \pm L\partial$ in the usual way: the problem is that we would like the contributions from the other terms with $i = 2, 3$ to vanish. To achieve this we will deform all the holonomy maps $\tilde{h}_{\rho\sigma}$, preserving the product relation Equation 7.12. We exploit the fact that $SU(2)$ is the 3-sphere. Choose a family of smooth maps,

$$\psi_r : SU(2) \to SU(2),$$

for $r \in (0, 1]$, such that

- each ψ_r is equivariant with respect to the adjoint action of $SU(2)$,
- ψ_1 is the identity,
- $\psi_r(-1) = -1$ for all r,
- when r is small ψ_r crushes the complement of the r-ball about -1 to $1 \in SU(2)$.

Now the frames we have chosen for the fibres of the flat bundles over the base point can be extended to frames for sufficiently nearby connections: that is, we can choose disjoint closed neighbourhoods $N_\sigma \subset \mathcal{B}^*_{\mathcal{Y}}$ of the flat connections σ and trivialisations of the base point fibration over these neighbourhoods. Given any instanton $\mathbf{A}$ in $M_{\rho\tau}$ there are a finite set of intervals $I_1, \dots, I_p$ in $\mathbf{R}$ defined by the condition that $t \in \bigcup I_p$ if and only if A_t lies in some N_σ for $\sigma \neq \rho, \tau$. So each interval I_j is associated to a flat connection σ_j. There is a similar half-infinite interval I_{p+1} associated to the connection τ. Write the intervals I_j for $j \leq p$ as $[a_j, b_j]$ and the interval I_{p+1} as $[a_{p+1}, \infty)$ with $a_1 < b_1 < a_2 < b_2 < \cdots < b_p < a_{p+1}$. Fix a small positive number δ and a function $r(T)$ with $r(T) = 1$ if T is small and $r(T) = \delta$ if T is large. Let T_j be the length $b_j - a_j$ of the interval I_j, and define the modified holonomy $H_{\rho\tau}(\mathbf{A})$ as follows. We parallel transport the chosen frame over the base point for ρ, i.e. at $-\infty$, to a frame over the point (a_1, z_0) in $\mathbf{R} \times Z$. This gives a frame which we can compare with the chosen frame for σ_1 using the identifications we have fixed, so we get a group element g_1, comparing the two frames. We next apply the map ψ_{r_1} to g_1, where $r_1 = r(T_1)$, to get $g_1' = \psi_{r_1}(g_1)$ We can think of g_1' as defining a frame for our bundle over $b_1 \times y_0$. Then we parallel transport this to $a_2 \times y_0$ and repeat the process. (For the final interval I_{p+1} we apply the map ψ_δ.) The upshot is that we define maps

$$H_{\rho,\tau} : \overline{M'_{\rho\tau}} \to SU(2)$$

for all ρ, τ such that $H_{\rho,\tau} = H_{\rho,\sigma} H_{\sigma\tau}$ on the relevant face of the boundary, but now having the property that they map the bulk of the moduli spaces to the identity. To say this more precisely, assume that the original holonomy maps $\tilde{h}_{\rho\sigma}$ are transverse to -1. (We leave as another exercise for the reader to modify the construction to avoid this assumption.) Then in particular the holonomy $\tilde{h}$ does not take the value -1 on any moduli space of dimension less than 3, and it follows that the modified holonomy is equal to the identity (for suitable values of the various parameters) near any boundary point associated to factorisations with $i = 2$ or 3. We take a representative α for $H^3(SU(2))$ and for connections ρ, σ of index difference 4 we define

$$n_{\rho\sigma} = \langle H^*_{\rho,\sigma}(\alpha), M'_{\rho\sigma} \rangle.$$

These are the matrix entries of a linear map $\tilde{U}$ and this *is* a chain map since the construction makes the contributions from the other faces of the compactification of the 5-dimensional moduli space equal to zero.

Proposition 7.7 *The chain map $\tilde{U}$ induces the cap product U on $HF_*(Z)$.*

First, we have seen that the cap product U can be induced by any representative for the cohomology class in $H^4(\mathcal{B}^*_Z)$. In particular we can take a connection Γ on the base point fibration $\mathcal{E}$ over $\mathcal{B}^*_Z$ and the Chern–Weil representative by a 4-form which this defines. This gives one chain map U say. Likewise we take the standard (translation-invariant) volume form on $SU(2)$ to define the cochain α, this gives another chain map $\tilde{U}$. Now our modified-holonomy construction defines a trivialisation of $\mathcal{E}$ over certain moduli spaces of instantons on the tube, viewed as subsets of $\mathcal{B}^*_Z$. In particular if ρ and τ have index difference 3 we get a trivialisation of $\mathcal{E}$ over $M'_{\rho\tau} \times \mathbf{R}$ which extends to the compactification in $\mathcal{B}^*_Z$. Then the Chern–Simons form of Γ in this trivialisation is a 3-form over the 3-manifold $M'_{\rho\tau} \times \mathbf{R}$ and we can integrate this to define a number $c_{\rho\tau}$. We view this collection of numbers as the matrix of a linear map $c : C_* \to C_{*-3}$ in the usual fashion. We claim that c gives a chain homotopy between U and $\tilde{U}$. Suppose ρ and σ have index difference 4. The modified holonomy does not give a trivialisation of $\mathcal{E}$ over $M'_{\rho\sigma} \times \mathbf{R}$ but we can divide this space into two pieces P^+, P^-, where P^- is the intersection with a small neighbourhood of σ, such that we do get trivialisations over the two pieces. We can just let

$$P^+ = \{(A,t) \in M'_{\rho\sigma} \times \mathbf{R} : t \leq \beta\}, \; P^- = \{(A,t) \in M'_{\rho\sigma} \times \mathbf{R} : t \geq \beta\},$$

for some large $\beta \in \mathbf{R}$. These trivialisations extend to the closures of $P^{\pm}$ in $\mathcal{B}_Z^*$. The transition function relating the two trivialisations over $M'_{\rho\sigma} \times \{\beta\} \cong M'_{\rho\sigma}$ is exactly our map $H_{\rho\sigma}$. Now

$$\int_{M'_{\rho\sigma}\times\mathbf{R}} \mathrm{Tr} F_\Gamma^2 = \int_{P^+} \mathrm{Tr} F_\Gamma^2 + \int_{P^-} \mathrm{Tr} F_\Gamma^2,$$

and on each piece $P^{\pm}$ we can write

$$\mathrm{Tr} F_\Gamma^2 = d\phi^{\pm}$$

say, where $\phi^{\pm}$ are the Chern–Simons forms in the two trivialisations. Over the intersection $M' \times \{\beta\}$ these two Chern–Simons forms differ by the pull-back $H_{\rho\sigma}^*(\alpha)$ of the volume form. If we apply Stokes' theorem to the two pieces $P^{\pm}$ we get

$$\int_{M'_{\rho\sigma}\times\mathbf{R}} \mathrm{Tr} F_\Gamma^2 = \int_{M'_{\rho\sigma}} H_{\rho\sigma}^*(\alpha) + \int_{\partial P^+} \phi^+ + \int_{\partial P^-} \phi^-,$$

where $\partial P^+, \partial P^-$ are the portions of the codimension-1 part of the boundary of $M'_{\rho\sigma} \times \mathbf{R}$ which lie in the closure of P^+, P^- respectively, counted with appropriate multiplicity. A little thought shows that the contribution from these latter terms is precisely given by the matrix entry of $\partial c \pm c\partial$, which establishes the chain homotopy relation.

7.3.3 The reducible connection

So far we have been working in the framework of non-trivial $U(2)$ bundles, where all flat connections are irreducible. We will now discuss how the preceding theory must be modified in the case when Y is a homology 3-sphere. We can carry through the same constructions to define maps

$$U : C_i \to C_{i-4}$$

by evaluating appropriate cochains. There is just one point of difference which occurs when $\langle\rho\rangle$ is in CF_1 and $\langle\sigma\rangle$ is in CF_{-4}, so we have a 4-dimensional reduced moduli space $M'_{\rho\sigma}$. The boundary of $M'_{\rho\sigma}$ contains a new contribution coming from factorisations through the trivial connection θ. For each pair of instantons A from ρ to θ and B from θ to ρ there is an end of the moduli space $M'_{\rho\sigma}$ modelled on the space of gluing parameters $SO(3)$. Clearly the holonomy map on this boundary component is a translate of the identity map of $SO(3)$, so it follows from the 3-dimensional description of the previous Subsection that this contributes $\pm 1/4$ when one evaluates the cochain. The number of pairs (A, B), counted with signs, is just the matrix entry of the composite $D_2 \circ D_1 : CF_1 \to CF_{-4}$, and one arrives at the following.

Proposition 7.8 *If Y is a homology 3-sphere the procedure above defines a map $U : C_* \to C_{*-4}$ such that*

$$\partial U - U\partial = -\tfrac{1}{4} D_2 \circ D_1$$

on CF_1 (and $\partial U = U\partial$ on the other chain groups).

Of course there are many choices involved in the definition of the map U of the above Proposition: we may need to make a generic perturbation to define the Floer chains C_* and we need to choose a representative cochain to define U. It is not so easy now to distil out the topologically invariant data. To do this we define a $(\mathbf{Z}/8)$-graded chain complex $(\widetilde{CF}_*, \tilde{\partial})$ with

$$\widetilde{CF}_i = C_i \oplus C_{i-3}$$

for $i \neq 0$, and

$$\widetilde{CF}_0 = C_0 \oplus \mathbf{Q} \oplus C_{-3}.$$

The differentials are given by the direct sum of three terms

- $\begin{pmatrix} \partial & U \\ 0 & \partial \end{pmatrix} : \widetilde{CF}_i \to \widetilde{CF}_{i-1}, \qquad (7.13)$
- $\frac{1}{2} D_1 : (C_1 \subset \widetilde{CF}_1) \to (\mathbf{Q} \subset \widetilde{CF}_0)$,
- $\frac{1}{2} D_2 : (\mathbf{Q} \subset \widetilde{CF}_0) \to (CF_{-4} \subset \widetilde{CF}_{-1})$.

Then one can check that the four equations

$$\begin{aligned} \partial^2 &= 0, && (7.14)\\ D_1\partial &= 0, && (7.15)\\ \partial D_2 &= 0, && (7.16)\\ \partial u - u\partial + \tfrac{1}{4} D_1 D_2 &= 0 && (7.17) \end{aligned}$$

we have established are equivalent to the single equation $\tilde{\partial}^2 = 0$.

We have a filtration

$$C_* \subset C_* \oplus \mathbf{Q} \subset \widetilde{CF}_*, \qquad (7.18)$$

so we have a filtered chain complex. Notice that the complex $CF_* \oplus \mathbf{Q}$ here is the same as $\underline{CF}_*$ and that the quotient

$$\widetilde{CF}_*/CF_*$$

can be identified with $\overline{CF}_*$. We have a *vector space* involution $\sigma : \widetilde{CF}_* \to \widetilde{CF}_*$ which maps $\mathbf{Q}$ to itself by the identity map and which is the canonical identification of the copy $CF_i \subset \widetilde{CF}_i$ with $CF_i \subset \widetilde{CF}_{i+3}$. Thus, to abstract the algebra of the set-up we make the following definition.

Definition 7.9 *An $(\mathcal{F}, \sigma)$-complex $\tilde{C}_*$ consists of the following data:*

- *a $(\mathbf{Z}/8)$-graded complex $\tilde{C}_*$ of rational vector spaces,*
- *a subcomplex $C_* \subset \tilde{C}_*$,*
- *an injection $\mathbf{Q} \to \tilde{C}_0$ such that*

$$C_* \subset C_* \oplus \mathbf{Q} \subset \tilde{C}_*$$

  defines a filtration of $\tilde{C}_*$,
- *an involution $\sigma : \tilde{C}_* \to \tilde{C}_*$ which is the identity map on $\mathbf{Q}$ and such that the restriction of σ to C_* is of degree 3 and induces an isomorphism from C_* to $\tilde{C}_*/C_* \oplus \mathbf{Q}$ and hence gives a canonical vector space isomorphism*

$$\tilde{C}_* = C_* \oplus \mathbf{Q} \oplus \sigma C_*.$$

  We require that the component of the differential mapping σC_* to σC_* should be $\sigma \partial \sigma^{-1}$, where ∂ is the differential in C_*, and we require that the component of the differential mapping $\mathbf{Q}$ to $\mathbf{Q}$ should vanish.

We have seen then that, making various choices, we can associate an $(\mathcal{F}, \sigma)$-complex to a homology sphere Y.

We now go on to consider maps between these complexes. So suppose that $\tilde{C}_*, \tilde{C}'_*$ are $(\mathcal{F}, \sigma)$-complexes as above. The space of all $\mathbf{Q}$-linear maps $\mathrm{Hom}(\tilde{C}_*, \tilde{C}'_*)$ has a standard grading and differential, making it into a $(\mathbf{Z}/8)$-graded complex. The degree-0 cycles in this complex are just the chain maps from $\tilde{C}_*$ to $\tilde{C}'_*$ and two maps differ by a boundary in the complex precisely when they are chain-homotopic. We let $\mathrm{Hom}_{\mathcal{F}}(C_*, C'_*)$ denote the maps $\lambda : \tilde{C}_* \to \tilde{C}'_*$ which

- respect the filtrations of $\tilde{C}_*, \tilde{C}_*$,
- have zero component mapping $\mathbf{Q} \subset \tilde{C}_*$ to $\mathbf{Q} \subset \tilde{C}'_*$,

- satisfy $\sigma' \circ \lambda = \lambda \circ \sigma$.

Such a map λ is determined by four components:

- $\lambda_1 : (C_* \subset \tilde{C}_*) \to (C'_* \subset \tilde{C}'_*)$;
- $\lambda_2 : (\sigma C_* \subset \tilde{C}_{*+3}) \to C'_* \subset \tilde{C}'_*)$;
- $\lambda_3 : (\sigma C_* \subset \tilde{C}_{*+3}) \to (\mathbf{Q} \subset \tilde{C}'_0)$;
- $\lambda_4 : (\mathbf{Q} \subset \tilde{C}_0) \to (C'_* \subset \tilde{C}'_*)$.

Thus, as a rational vector space,

$$\begin{aligned}\mathrm{Hom}_{\mathcal{F}}(\tilde{C}_*, \tilde{C}'_*) &= \mathrm{Hom}(C_*, C'_*) \oplus \mathrm{Hom}(C_*, C'_*) \\ &\oplus \mathrm{Hom}(C_*, \mathbf{Q}) \oplus \mathrm{Hom}(\mathbf{Q}, C'_*).\end{aligned}$$

Now we have an obvious copy of $\mathbf{Q}$ in $\mathrm{Hom}(\tilde{C}_*, \tilde{C}'_*)$, and a direct sum decomposition of $\mathrm{Hom}(\tilde{C}_*, \tilde{C}'_*)$ as $\mathbf{Q}$, four copies of $\mathrm{Hom}(C_*, C'_*)$, two copies of $\mathrm{Hom}(C_*, \mathbf{Q})$ and two copies of $\mathrm{Hom}(\mathbf{Q}, C'_*)$. So we have a vector space isomorphism

$$\mathrm{Hom}(\tilde{C}_*, \tilde{C}'_*) = \mathrm{Hom}_{\mathcal{F}}(\tilde{C}_*, \tilde{C}'_*) \oplus \mathbf{Q} \oplus \tau(\mathrm{Hom}_{\mathcal{F}}(\tilde{C}_*, \tilde{C}'_*)), \qquad (7.19)$$

for a suitable injection $\tau : \mathrm{Hom}_{\mathcal{F}} \to \mathrm{Hom}$.

Lemma 7.10 $\mathrm{Hom}_{\mathcal{F}}(C_*, C'_*) \subset \mathrm{Hom}_{\mathcal{F}} \oplus \mathbf{Q}$ *are subcomplexes of* $\mathrm{Hom}(C_*, C'_*)$ *and, for a suitable choice of the map* τ*, the decomposition Equation 7.19 makes* $\mathrm{Hom}(\tilde{C}_*, \tilde{C}'_*)$ *into an* $(\mathcal{F}, \sigma)$*-complex.*

We leave the proof to the reader. Now define a category $\mathcal{C}$ as follows. An object of $\mathcal{C}$ is an $(\mathcal{F}, \sigma)$-complex and a morphism is an equivalence class of degree-0 chain maps in $\mathrm{Hom}_{\mathcal{F}} \oplus \mathbf{Q}$ up to chain homotopies by elements of $\mathrm{Hom}_{\mathcal{F}} \oplus \mathbf{Q}$ (that is, a homology class in the $(\mathrm{Hom}_{\mathcal{F}} \oplus \mathbf{Q})$-complex). We define another category $\mathcal{H}$ whose objects are oriented homology 3-spheres with base points, and whose morphisms are oriented homology cobordisms, with a choice of homotopy class of path joining the base points on the boundary.

Then we have

Theorem 7.11 *The construction above defines a functor from the category* $\mathcal{H}$ *to the category* $\mathcal{C}$*.*

We will outline the proof of this result.

Let Y, Y' be two oriented homology 3-spheres, with base points, and, making appropriate choices, let

$$\widetilde{CF}_* = (C_*, \partial, U, D_1, D_2), \ \widetilde{CF}'_* = (C'_*, \partial', U', D'_1, D'_2)$$

be the data defined as above for the two manifolds

Now let W be a homology cobordism from Y to Y' and choose a path joining the base points on the boundary. We need to show first that, with suitable choices, W defines a *chain map* $\underline{\lambda}$ in $\mathrm{Hom}_{\mathcal{F}}$. To do this we need to define maps $\lambda_1, \dots, \lambda_4$ as above and a further component λ_5 in $\mathbf{Q}$. These are given as follows.

- The map λ_1 is the ordinary map on the Floer chains, as defined in Chapter 5, using 0-dimensional moduli spaces of instantons over W.
- The map λ_2 is defined by 3-dimensional moduli spaces of instantons over W. We define a modified holonomy map on these moduli spaces, much as above, and evaluate the pull-back of a 3-cochain to get the matrix entries of the maps.
- The map λ_3 is defined by counting instantons over W with limits an irreducible connection ρ over Y and the trivial connection θ' over Y'.
- The map λ_4 is defined by counting instantons over W with limits the trivial connection θ over Y and an irreducible connection ρ' over Y'.
- The map λ_5 is the identity map from $\mathbf{Q}$ to $\mathbf{Q}$ (which corresponds geometrically to the trivial connection over W).

To obtain the relations satisfied by these maps we consider four kinds of moduli spaces of instantons over W:

- One-dimensional moduli spaces of instantons with irreducible limits at either end. Counting the boundary components gives the relations

$$\partial' \lambda_1 = \lambda_1 \partial$$

  which we have seen already in Chapter 5.
- Four-dimensional moduli spaces of instantons with irreducible limits at either end. We extend the modified holonomy map, and hence the 3-cochain, over these moduli spaces. The fact that the evaluation on the boundary gives 0 yields the relation

$$\partial' \lambda_2 - \lambda_2 \partial = \lambda_4 D_1 - D'_1 \lambda_3.$$

- One-dimensional moduli spaces with an reducible limit over Y and the trivial limit θ' over Y'. The boundary of these gives

$$D_1 = \lambda_5 D_1 = D_1' \lambda_1.$$

- One-dimensional moduli spaces with the trivial limit θ over Y and an irreducible limit over Y'. The boundary of these gives

$$\lambda_1 D_2 = D_2' = D_2' \lambda_5.$$

These relations are precisely what is needed to show that the map $\underline{\lambda}$ is a chain map.

To complete the proof we need to show two further things: first, that changing the various choices used to define $\underline{\lambda}$ changes the map by an appropriate chain homotopy; second, that if we have another cobordism from Y' to Y'' say then, with suitable choices, the chain maps of the composite cobordism are the composite of the two chain maps as above. Both of these follow just the same principles as in the case of the ordinary Floer groups and we leave details to the reader.

7.3.4 Equivariant theory

There are a number of variants of the Floer groups that we can define which build in the information from the trivial connection and cup product. The point of Theorem 7.11 is that this discussion now becomes purely a matter of algebra: any functor from the category $\mathcal{C}$ will give rise to a topological invariant of homology 3-spheres. The first construction we consider runs as follows. Given an object in our category $\mathcal{C}$ – represented by $(C_*, \partial, D_1, D_2, U)$ – let $\mathbf{Q}[\![y]\!]$ denote the ring of formal power series in an indeterminate y and define a $(\mathbf{Z}/8)$-graded vector space

$$\overline{\overline{CF}}_* = \overline{\overline{CF}}_*(Y) = C_*(Y) \oplus \mathbf{Q}[\![y]\!],$$

where the generator y is thought of as having degree 4, so for example

$$\overline{\overline{CF}}_0 = C_0 + \mathbf{Q} + \mathbf{Q}(y^2) + \mathbf{Q}(y^4) + \cdots.$$

We define a differential $\overline{\overline{\partial}}$ to be the sum

$$\overline{\overline{\partial}} = \partial + D_1 + yD_1 \circ U + \frac{y^2}{2!} D_1 \circ U \circ U + \frac{y^3}{3!} D_1 \circ U \circ U \circ U + \cdots. \quad (7.20)$$

Here D_1 is the familiar map from C_1 to $\mathbf{Q} \subset \mathbf{Q}[\![y]\!]$ and 'y' denotes the operation of multiplication by y in $\mathbf{Q}[\![y]\!]$. In a more condensed notation we can write

$$\overline{\overline{\partial}} = \partial + D_1 e^{yU}.$$

We also define a degree-(-4) map

$$\overline{\overline{U}} : \overline{\overline{CF}}_* \to \overline{\overline{CF}}_*,$$

by

$$\overline{\overline{U}} = U + \tfrac{1}{4} D_2 + \frac{\partial}{\partial y} \tag{7.21}$$

where D_2 means, strictly, the composite of the map we have considered before from $\mathbf{Q}$ to C_{-4} with the 'evaluation map' $\mathbf{Q}[\![y]\!] \to \mathbf{Q}$ setting y to 0. For example

$$\overline{\overline{U}}(y^2 + y + 3) = \frac{3}{4} D_2(1) + 2y + 1 \in C_{-4} \oplus \mathbf{Q}[\![y]\!].$$

It is an exercise in algebra to check that $(\overline{\overline{CF}}_*, \overline{\overline{\partial}})$ is a chain complex and $\overline{\overline{U}}$ is a chain map, so we get a cohomology group

$$\overline{\overline{HF}}_*(Y)$$

with an endomorphism which we will again denote by

$$U : \overline{\overline{HF}}_* \to \overline{\overline{HF}}_*.$$

In other words HF_* is a module over the polynomial ring $\mathbf{Q}[u]$ in one variable. Similarly one checks that, up to the obvious equivalence, this depends only on the chain homotopy class of the original data, so in particular we get an invariant of homology 3-spheres. This construction can be seen as a generalisation of that of the $\overline{HF}_*$. Indeed we can make a sequence of such constructions, for each $p \geq 0$, by following the same procedure as above but setting $y^p = 0$. In each case we get a chain complex and homology group. The case $p = 1$ gives $\overline{HF}_*$ whereas $\overline{\overline{HF}}_*$ is in a natural way the limit of these constructions as $p \to \infty$. It is only in this limit however that we can define the product map on homology.

Clearly the obvious inclusions of chain complexes define maps on homology

$$\overline{\overline{HF}}_* \to \overline{HF}_* \to HF_*. \tag{7.22}$$

We have already discussed in Section 7.1 how the basic 4-manifold invariants can be lifted from HF_* to $\overline{HF}_*$. We can now take this further and try to lift our invariants to $\overline{\overline{HF}}_*$. Let X_1 be a 4-manifold with $b^+ > 1$ and boundary Y. Recall that $A(X_1)$ is the ring generated formally by the homology of X_1. In particular $A(X_1)$ is a module over the polynomial ring $\mathbf{Q}[u]$ on one variable, corresponding to one half the generator of $H_0(X_1)$.

Proposition 7.12 *There is a way to define a* $\mathbf{Q}[u]$*-module homomorphism* $\overline{\overline{\psi}} : A(X_1) \to \overline{\overline{HF}}_*(Y)$ *which lifts the previous linear map of vector spaces* $\overline{\psi} : A(X_1) \to \overline{HF}_*(Y)$.

Of course, as usual, this map $\overline{\overline{\psi}}$ will be an invariant of the smooth manifold-with-boundary (X_1, Y).

Proposition 7.12 may be compared with the corresponding assertion working with non-trivial bundles over a 3-manifold Z with non-trivial H^1. In that case the ordinary Floer group already has a cap product, i.e. is a $\mathbf{Q}[u]$ module, and the fact that the invariant of a 4-manifold with boundary is a module homomorphism follows immediately from the basic theory. In the case of a homology sphere the construction of $\overline{\overline{HF}}_*$ is just what is needed to repair the damage done by the trivial connection. (A slightly different approach is discussed in Section 8.1 below.)

The proof of Proposition 7.12 brings in some new ideas. First, the other homology classes in X_1 will play no real role in the discussion, so we may as well concentrate on the copy of $\mathbf{Q}[u]$ in $A(X_1)$, and since our map is to be a $\mathbf{Q}[u]$-homomorphism it suffices to define $\overline{\overline{\psi}}$ on the identity. As usual, this will be defined via a chain in the $\overline{\overline{CF}}$-complex. This chain has a component in $CF_* \subset \overline{\overline{CF}}_*$, which will just be the familiar Floer chain defined by 0-dimensional moduli spaces with irreducible limits. The new feature is that we have to define a component

$$\sum_j n_j y^j,$$

say, in $\mathbf{Q}[\![y]\!]$. The basic idea is that n_j is given by evaluating $\mu(\frac{1}{2}\text{pt.})^j$ on a $4j$-dimensional moduli space of instantons over X_1 with the trivial limit θ. (Thus, depending on the value of $(1 - b_1 + b_+)$ modulo 8, either all the n_j will be zero – if there is no moduli space of dimension $4j$ – or all the n_j will be zero for j odd or for j even.) At this level, the chain data appears to be exactly the same as we have used already in defining $\overline{\psi}(U^j)$ in $\overline{HF}_*(Y)$. However, in fact Proposition 7.12 is a genuine refinement of the $\overline{HF}$ theory, and the class $\overline{\overline{\psi}}$ contains more information. This comes about because we use particular kinds of representatives for $\mu(\text{pt.})$ to define the numbers n_j.

For simplicity we will just consider the definition of the number n_1, in the case when there is a 4-dimensional moduli space M_θ of instantons over X_1 with the trivial limit. We fix some representative cochain S for the 4-dimensional cohomology class over $\mathcal{B}_Y^*$, let us say by a codimension-4 submanifold. We choose this to be in general position with respect to all the relevant finite-dimensional moduli spaces, in particular we arrange that no irreducible flat connection lies in the support of S. Now for any $L > 0$ we can restrict a connection over X_1 to the copy $Y_L = Y \times \{L\} \subset Y \times (0, \infty) \subset X_1$ of Y. Pulling back by this restriction map we get a 4-dimensional cochain S_L, or more precisely a codimension-4 subset, over M_θ. For fixed, generic, L we can evaluate this (i.e. count the subset) on the moduli space to get a number $n_1(L)$. This is just a part of the standard theory. Similarly, the standard theory tells us that if we change L, or any other parameter, the number $n_1(L)$ will change by $D_1(\lambda)$ for some λ in CF_1. The new feature here is that we show more:

- for sufficiently large L the numbers $n_1(L)$ do not vary with L, i.e. $n_1(L)$ has a limit, n_1 say, as L tends to infinity;
- under a compactly supported variation of metric on X_1 the number n_1 changes by $D_1(\lambda)$ for some $\lambda \in CF_5$.

To prove the first part, consider how the number $n_1(L)$ might change as L increases through some critical value L_∞. This can only happen if there are a sequence $L_i \to L_\infty$ and connections $A_i \in M_\theta$ such that A_i lies in S_{L_i} but the A_i do not converge in M_θ as i tends to infinity. Standard arguments rule out any 'bubbling' and one sees that the only thing that can happen is that the A_i are chain-convergent, given for large i by gluing an instanton A over X_1 with irreducible limit ρ to an instanton I over the tube running from ρ to θ. (Again, familiar arguments rule out longer chains.) Let M_ρ be the moduli space of instantons over X_1 containing A. The additivity of the index shows that the dimension of M_ρ is at most 3. Now we may suppose that L_∞ is chosen large enough so that the restriction of any connection in M_ρ to the segment $Y \times (L_\infty - 1, L_\infty + 1)$ is either close to ρ or well approximated by an instanton J on the tube running from some other flat connection σ to ρ. The connection A cannot be of the first kind, since ρ does not lie in the support of S. In the second case, the index of J is also at most 3 and the moduli space M_ρ in a neighbourhood of A is modelled on a product of the moduli space $M_{\sigma\rho}$ of instantons on the tube, containing J, and another factor M_σ, of instantons over X_1. It follows then that, near the connections A_i, the restriction map from M_θ to Y_L, for L near L_∞, is close to the restriction map from $M_{\sigma\rho}$ to Y_L. But $M_{\sigma\rho}$ contains a translation parameter, which is obviously equivalent to changes in L. We suppose by general position that the restriction of $M_{\rho\sigma}$ to Y_{L_∞} – which yields a set of dimension at most 3 in $\mathcal{B}^*_Y$ – does not meet the support of S. Then it follows that the restriction of $M_{\rho\sigma}$ to any Y_L does not meet S and this gives the desired contradiction.

For the second part, we carry through a similar discussion with L fixed and sufficiently large and a one-parameter family of metrics on X_1, all fixed over the tubular end. The same argument shows that the only way a change could occur is when a sequence A_i is chain-convergent to a connection A in M_ρ glued to an instanton I running from ρ to θ, and where in turn the moduli space near A is modelled by gluing a moduli space of instantons $M_{\sigma\rho}$ over the tube to connections B over X_1 with limit σ. Our dimension counting now shifts by 1, since in the one-parameter family of metrics we expect to meet solutions B with index -1, and this allows $M_{\sigma\rho}$ to have dimension 4. Conversely, the usual gluing arguments show that for any connections B, I, J as above we get a contribution to the change in n_1. But if we use this representative S to define the cap product map U on $C_*(Y)$, the number of pairs I, J counted with signs just gives $D_1U(\lambda)$, where λ is the Floer chain defined by counting the solutions B of index -1 in the one-parameter family. This establishes the second item above. Of course we see more: the change in n_1 by $D_1U(\lambda)$ is accompanied in a change in the ordinary Floer chain by $\partial\lambda$, which just says that the class in the complex $\overline{\overline{CF}}$ does not change.

The proof of Proposition 7.12 in the general case is similar. We define n_i by choosing large parameters

$$0 \ll L_1 \ll L_2 \ll \cdots \ll L_i,$$

and pull-back representatives S_i by the restriction map to the Y_{L_i}. The same analysis shows that the intersection of M_θ with all the S_i defines a number n_i which is independent of the L_α, provided these are all sufficiently large, and that a change under a one-parameter family of metrics is associated to configurations of connections corresponding to the composite map D_1u^i. The fact that the map so defined both is a $\mathbf{Q}[u]$-homomorphism and lifts $\overline{\psi}$ is a reflection of the compatibility of the chain data noted above. One can go on to show that the constructions are independent of the choices made over Y.

Now, in much the same way as we defined two groups $\overline{HF}_*, \underline{HF}_*$, we can define another group $\underline{\underline{HF}}_*$. Here we take a formal variable x and let

$$\underline{\underline{CF}}_* = C_* \oplus \mathbf{Q}[x],$$

where $\mathbf{Q}x^p \subset \mathbf{Q}[x]$ is assigned degree $4p-3$. Notice that in this case we are taking polynomials, rather than formal power series. The differential has the ordinary component ∂ on C_* and an additional component mapping $\mathbf{Q}[x]$ to CF_* given by

$$\underline{\underline{\partial}}(f(x)) = f(u)D_2(1).$$

One checks this is a chain complex and we define $\underline{\underline{HF}}_*(Y)$ to be its homology. There is a chain map $\underline{\underline{U}}$ on $\underline{\underline{CF}}_*$ given by

$$\underline{\underline{U}}(\rho + f(x)) = u\rho + \tfrac{1}{4}D_1\rho + xf(x),$$

and this makes $\underline{\underline{HF}}_*$ a $\mathbf{Q}[u]$-module. This construction is the adjoint of the previous one, in the sense that the obvious pairing

$$\underline{\underline{CF}}_*(Y) \otimes \overline{\overline{CF}}_*(\overline{Y}) \to \mathbf{Q}$$

induces a pairing

$$\underline{\underline{HF}}_*(Y) \otimes \overline{\overline{HF}}_*(\overline{Y}) \to \mathbf{Q} \tag{7.23}$$

under which

$$\underline{\underline{HF}}_*(Y) = \mathrm{Hom}(\overline{\overline{HF}}_*(\overline{Y}), \mathbf{Q}).$$

The products are adjoint with respect to the pairing. We also have natural maps

$$HF_*(Y) \to \underline{HF}_*(Y) \to \underline{\underline{HF}}_*(Y),$$

adjoint to Formula 7.22.

As usual, finite-dimensional analogues shed light on the constructions of the preceding paragraphs. Suppose V is some compact finite-dimensional manifold and a compact group G acts on V. Let $EG \to BG$ be a universal G bundle – so EG is a weakly contractible space on which G acts freely. The *equivariant cohomology* $H^*_G(V)$ is defined to be the ordinary cohomology of the 'homotopy quotient'

$$V_G = \frac{V \times EG}{G},$$

where G acts on both factors. We may also consider the equivariant homology $H_*(V_G)$. There is a projection map from V_G to BG which induces a map from $H^*(BG)$ to the cohomology ring $H^*(V_G)$, and this makes $H^*_G(V)$ into a module over $H^*(BG)$. At one extreme, if G acts freely on V then V_G is weak homotopy equivalent to the ordinary quotient space V/G, so $H^*_G(V) = H^*(V/G)$ and the $H^*(BG)$ module structure is just given by the products with the characteristic classes of the G bundle $V \to V/G$. At the other extreme, if G acts trivially the homotopy quotient is the product $V \times BG$ and the module structure is the product in $H^*(BG)$ on the second factor.

Now suppose that $f : V \to \mathbf{R}$ is a G-invariant function on V. We seek a generalisation of the Morse description of ordinary cohomology to the equivariant case. The function f induces a function $\tilde{f} : V_G \to \mathbf{R}$. The fact that the spaces EG and V_G are not finite-dimensional manifolds does not really matter much. One can either work with finite-dimensional approximations or adapt the theory directly to an infinite-dimensional situation. Now the function $\tilde{f}$ cannot be a Morse function. Consider the projection map π from V_G to the quotient V/G. For each point $[x]$ in V/G we choose a representative x in V and look at the stabiliser $\Gamma_x \subset G$ in G. Up to conjugation this is independent of the representative chosen. The fibre $\pi^{-1}(x)$ is EG/Γ_x which is homotopy-equivalent to the classifying space $B\Gamma_x$. By construction, the function $\tilde{f}$ is constant along the fibres of π, so its critical set is a union of fibres. We assume that the function f is an 'equivariant Morse function', which means that the critical set C of $\tilde{f}$ is a finite union of fibres of π,

$$C = \pi^{-1}(x_0) \cup \cdots \cup \pi^{-1}(x_n),$$

and the Hessian of the function is non-degenerate transverse to the fibres. We are thus in the general setting of 'Morse–Bott' theory. To each component $C_i = \pi^{-1}(x_i)$ of the critical set we assign an index μ_i – the dimension of a maximal negative subspace of the Hessian in the normal directions. We let C^p be the union of the components with index p. On these components we take a suitable model (Ω^*, d) for the ordinary cohomology cochain complex: for example (if we are working with real co-efficients) we could use the differential forms and exterior derivative. The result we want is the following, which is a more or less standard part of Morse–Bott theory. (We refer to [5] for more details.)

Proposition 7.13 *There is a way to define a filtered complex with groups*

$$E^{pq} = \Omega^q(C^p),$$

and differential $\underline{d} = \sum_{r\geq 0} d_r$,

$$d_r : E^{pq} \to E^{p+r\; q-r+1},$$

where $d_0 = d : \Omega^*(C^p) \to \Omega^*(C^p)$, *and whose cohomology is the equivariant cohomology* $H^*(V_G)$.

The higher differentials d_r can be defined using the manifolds of gradient flow lines $\mathcal{M}(C_i, C_j)$ between critical sets. We have end point maps

$$e_- : \mathcal{M}(C_i, C_j) \to C_i, \quad e_+ : \mathcal{M}(C_i, C_j) \to C_j,$$

and the differentials are the composites $(e_-)_* e_+^*$ of the pull-back and push-forward (integration over fibres) maps on forms. This description leads to an important general restriction on the differentials, reflecting the symmetry. We can start with a G-invariant Riemannian metric on V and then take the product metric on $V \times EG$. Let $\mathcal{N}$ be a moduli space of gradient lines on V: clearly G acts on $\mathcal{N}$. The gradient lines on $V \times EG$ are just $\mathcal{N} \times EG$, and hence we see that the corresponding moduli space of gradient lines $\mathcal{M}$ in V_G is itself the homotopy quotient $\mathcal{M} = \mathcal{N}_G = \mathcal{N} \times_G EG$. Now consider the end point maps

$$\mathcal{M} \to C_i, \quad \mathcal{M} \to C_j.$$

Suppose that Γ_i is trivial – so the G-action on $\mathcal{M}$ is free – but Γ_j is non-trivial. Then the fibres of the end point map e_+ contain non-trivial Γ orbits which map non-trivially under e_-. It follows that the composite of push-forward and pull-back in this situation is trivial. Thus the part of the differential mapping $H^*(C_j) = H^*(\text{Point})$ to $H^*(C_i) = H^*(B\Gamma)$ vanishes.

Proposition 7.13 implies that there is a spectral sequence converging to the equivariant cohomology whose E_1 term is

$$E_1^{pq} = H^q(C^p).$$

In other words E_1^{**} is the sum

$$\bigoplus H^*(B\Gamma_{x_i}),$$

but with the grading on $H^*(B\Gamma_{x_i})$ shifted by μ_i. Now, to be concrete, suppose we have an example where the stabilisers Γ_{x_i} are trivial for all $i > 0$, whereas $\Gamma_{x_0} = SO(3)$. Working with rational cohomology, $H^*(BSO(3))$ is the polynomial ring generated by one element x in H^4. Then the E_1 term is $C^* \oplus \mathbf{R}[x]$, (where C^* is generated by isolated critical points) and, taking account of the vanishing phenomenon above, the differentials are

$$d_1 : C^* \to C^{*+1}, \quad d_{4j+1} : \mathbf{R} \to C^{a+4j},$$

for some fixed a. Explicitly d_1 is given by counting gradient lines, i.e. points in $\mathcal{N}/G$, and d_{4j+1} is given by evaluating the pull-back of x^j under the end point map on a moduli space $\mathcal{N}/G$. We arrive then at a complex which is almost an exact analogue of the one we have used to construct $\underline{\underline{HF}}_*(Y)$. The only differences are that, first, our gradings are reversed, i.e. the Floer differential decreases degree. This is simply a matter of book-keeping and convention. Second, in the Floer case we defined the differential analogous to d_{4j+1} by using the jth power of the product map u. But, much as we have seen in the proof of Proposition 7.12, these give the same complex, up to chain homotopy.

The conclusion of this discussion is that we can regard the group $\underline{\underline{HF}}_*(Y)$, with its $\mathbf{Q}[u]$-module structure, as the Floer analogue of the equivariant *cohomology* of the space of connections modulo the gauge group. Likewise we can regard the group $\overline{\overline{HF}}_*(Y)$ as the analogue of the equivariant *homology*, using the duality Formula 7.23. We can apply the same ideas directly to the chain complex $\widetilde{CF}_*(Y)$. The simplest invariant we can form from this is its homology $\widetilde{HF}_*(Y)$. This can be interpreted in the above spirit as the Floer analogue of the cohomology of the space $\tilde{B}_Y$ of 'framed' connections over Y.

7.3.5 Limitations of existing theory

So far the theory has unrolled in a smooth, perhaps anodyne, fashion. We have been developing more and more elaborate algebraic gadgets but ones which encode the same basic geometrical operations: evaluating cohomology classes on moduli spaces and their boundaries. Moreover the Floer theory has followed very closely the analogous constructions in ordinary, finite-dimensional, homology theory. In essence this is because the moduli spaces we have been dealing with behave as though they were compact, except for the familiar 'factorisation' phenomenon. But now we meet a problem in which the other non-compactness mechanism – bubbling at points – plays a real role. Suppose that X_1 is a manifold with boundary Y, and with $b_1(X_1) = b^+(X_1) = 0$. Recall that in Subsection 7.2.2 we defined an invariant $\underline{\Psi}(X_1) \in \underline{HF}_*(Y)$ – the extra component in the differential being just what was needed to take account of the trivial connection when considering 0-dimensional moduli spaces over X_1. One might hope that this extends to a map from $A(X_1)$ to $\underline{\underline{HF}}_*(Y)$, performing a similar task for the higher-dimensional moduli spaces. We proceed in this direction by taking the familiar map ψ : $A(X_1) \to C_*(Y)$ into the chain groups, evaluating classes on moduli spaces of instantons with irreducible flat limits. It will be useful in this Subsection to work with $U(2)$ bundles, so we specify a first Chern class $v_1 \in H^2(X_1)$, but we shall not always make this explicit in our notation. For each $\lambda \in A(X_1)$ we would like to define an element $f_\lambda(x)$ of $\mathbf{Q}[x]$ so that

$$\underline{\underline{\partial}}(\psi(\lambda) + f_\lambda) = \partial(\psi(p)) + f_\lambda(u)D_2(1)$$

vanishes. Suppose, for example, that $\lambda = \alpha^d$ for some $\alpha \in H_2(X_1)$ and that $\psi(\lambda)$ lies in CF_{-3}, so $\partial(\psi(\lambda))$ is in CF_{-4}. Let $\langle\sigma\rangle$ be a generator of CF_{-4}. We expect that the required identity should follow by considering the ends of a $(2d + 1)$-dimensional moduli space M_σ of instantons with limit σ. That is, we want to exploit the formula

$$\langle \mu(\alpha)^d, \partial M_\sigma \rangle = 0 \qquad (7.24)$$

where, strictly, ∂M_σ is the boundary of a suitable large compact subset of M_σ. The $\langle\sigma\rangle$ component of $\partial\psi(\alpha^d)$ gives, in the familiar way, the contribution from ends of this moduli space arising from factorisations through irreducible flat connections over Y. We would like to define f_{α^d} to take account of factorisations through the trivial connection. We write $M_{\theta\sigma}$ for the moduli space of instantons on the tube running from the trivial connection θ to σ. In a familiar way again, we focus attention on the abelian reducible instantons over X_1, corresponding to cohomology classes $e \in H^2(X_1)$ with $e = v_1 \bmod 2$. Given such a class e we can consider the part of the end of M_σ obtained by gluing the reducible solution over X_1 to an instanton running from θ to σ over the tube. If $e \cdot e = -r$, our dimension formulae tell us that

$$\dim M_{\theta\sigma} = 2(d - r) + 1,$$

since the moduli space of instantons over X_1 containing the given reduction has dimension $2r - 3$. In particular $d \geq r$. We can analyse this situation using similar techniques to the 'blow-up trick' described in Section 6.3. The local model for this part of the end of M_σ is a $\mathbf{C}^{r-1}$ bundle over the total space of an S^2 bundle over $M_{\theta\sigma}$. The $\mathbf{C}^{r-1}$ fibres are easily dealt with: we can choose $(r - 1)$ standard representives for μ_α (as codimension-2 submanifolds) whose intersection (counted with multiplicity) is $(\alpha \cdot e)^{r-1}$ times the zero section of the bundle. Thus evaluating $\mu(\alpha)^d$ on this part of the end of the moduli space comes down, ignoring issues of compactness, to evaluating $\mu(\alpha)^{d-r+1}$ on the total space E of an S^2 bundle over the $2(d-r)$-dimensional reduced moduli space $M'_{\theta\sigma} = M_{\theta\sigma}/\mathbf{R}$. Now the bundle E is the sphere bundle associated to an $SO(3)$ bundle over $M_{\theta\sigma}$, which is just the usual base point fibration. The restriction of the cohomology class $\mu(\alpha)$ is $2(\alpha \cdot e)$ times the standard generator h of $H^2(E)$ whose square $h^2 \in H^4(E)$ is the lift of four times the Pontryagin class of the bundle, which is just -2 times our standard 4-dimensional cohomology class U. Finally there is an orientation factor, as in Proposition 7.3, which brings in a term $(-1)^{\frac{1}{2}(d+e\cdot e)}\sigma(e)$ where $\sigma(e) = +1$ if the square of $\frac{1}{2}(e - v_1)$ is even and $\sigma(e) = -1$ otherwise. Putting this all together, we expect that the contribution to $\langle \partial M_\sigma, \mu(\alpha^d)\rangle$ from this part of the end is

$$2^{3(d-r)/2+1}(-1)^{\frac{1}{2}(d+e\cdot e)}\sigma(e)(\alpha \cdot e)^d \langle U^{\frac{1}{2}(d-r)}, M_{\theta\sigma}/\mathbf{R}\rangle.$$

(This is to be interpreted as zero if $d - r$ is odd.) In turn we can relate the evaluation of powers of U to the product map on C_* by moving points apart along the tube, much as in the proof of Proposition 7.12. Taking account of all the reductions, this discussion suggests finally that we should define

$$f_{\alpha^d}(x) = \sum 2^{3(d+e\cdot e)/2}(-1)^{\frac{1}{2}(d+e\cdot e)}\sigma(e)(\alpha \cdot e)^d x^{(d+e\cdot e)/2}, \qquad (7.25)$$

where e runs over the integral classes with $d + e \cdot e$ a non-negative even integer, and with $e = v_1 \bmod 2$. Now the main point is that this 'naive' answer is *not correct* in general, although it is correct provided the moduli spaces which arise have sufficiently low dimension. Our notation above ignores the fact that there are moduli spaces of different dimensions over the tube, with the same flat limits. The difficulty occurs once one has to consider a moduli space $M^j_{\theta\sigma}$, say, of dimension $j > 8$. There is then another moduli space $M^{j-8}_{\theta\sigma}$ with these same flat limits and dimension $j - 8$. We know that there are additional parts of the boundary of $M^j_{\theta\sigma}$ corresponding to bubbling on the tube. Equally there is another part of the boundary of our original moduli space M_σ made up of the following configurations. Take the reducible solution over X_1 and glue in a small instanton at a point q in X_1, then glue this to a connection represented by a point of $M^{j-8}_{\theta\sigma}$. These regions in the end of the moduli space do make a contribution to the pairing Equation 7.24, but this is not taken into account by the putative Formula 7.25. (The two regions intersect, when the point q moves down the tubular end of X_1.) As the moduli space dimensions increase one has to consider more and more complicated bubbling over many points. The conclusion of the discussion above is that, lacking a close analysis of these extra contributions, we do not know how to define a map from $A(X_1)$ to $\underline{\underline{HF}}_*(Y)$ with the desired properties. We will say more about this in the next Chapter. However, the tools we have do allow us to define such a map on a subset of $A(X_1)$. Let

$$A^\nu(X_1) \subset A(X_1)$$

denote the elements of $A(X_1)$ of total degree at most 6. (Recall that we are assuming $b_1(X_1) = 0$, so there are only even-dimensional classes in $A(X_1)$.) Thus $A^\nu(X_1)$ is spanned by products

$$\alpha^d u^p$$

with $p = 1$ and $d \leq 1$ or $p = 0$ and $d \leq 3$. We define

$$\underline{\underline{\psi}} : A^\nu(X_1) \to \underline{\underline{HF}}(Y) \tag{7.26}$$

following the procedure outlined above. Thus, at the chain level, for each basis element $\alpha^d U^p$ we take the ordinary Floer chain in $CF_*(Y)$ and a $\mathbf{Q}[x]$ component

$$f_{\alpha^d U^p}(x) = \sum 2^{3(d+e\cdot e)/2} \sigma(e)(\alpha \cdot e)^d x^{\frac{1}{2}(d+e\cdot e)+p}, \tag{7.27}$$

where the sum runs over integral classes with $-e\cdot e \leq d-2p$, with $d+e\cdot e$ even and with $e = v_1 \bmod 2$. (Notice that these classes come in pairs, corresponding to the same geometrical reduction, and we have adjusted the factor of 2 accordingly.) Then one can show without difficulty that this gives a cycle in $\underline{\underline{CF}}_*$ and the homology class is independent of choices, so we get a map as in Formula 7.26. We also have a partial gluing formula.

Proposition 7.14 *If X is a closed 4-manifold decomposed as a sum $X_1 \cup_Y X_2$ with $b^+(X_1) = 0$ and $b^+(X_2) > 1$ then the invariants of X which lie in the image of $A^\nu(X_1) \otimes A(X_2)$ can be computed from the relative invariants*

$$A^\nu(X_1) \to \underline{\underline{HF}}_*(Y), \quad A(X_2) \to \overline{\overline{HF}}_*(\overline{Y})$$

of the two pieces via the pairing Formula 7.23.

The proof uses familiar techniques and is left to the reader.

We encounter similar difficulties when we seek to generalise Proposition 7.3. If we have a moduli space M_θ over the manifold X_1 with the trivial limit θ and of dimension $2d + 4p + 1$ we can assert that

$$\langle \mu(\alpha)^d U^p, \partial M_\theta \rangle = 0, \tag{7.28}$$

where again ∂M_θ denotes the boundary of a suitable large compact piece of the irreducible part of the moduli space. There are two familiar contributions. The first comes from factorisations through irreducible flat connections ρ with $\langle \rho \rangle \in CF_1$. The contribution from these is $D_1(\rho)$ times the pairing between $\mu(\alpha^d)U^p$ and the $(2d+4p)$-dimensional moduli space M_ρ. The second comes from abelian reductions. Write

$$c_{\alpha^d U^p} = 2^{-3p} \sum \sigma(e)(\alpha \cdot e)^d, \tag{7.29}$$

where the sum runs over classes e with $-e \cdot e = d+2p$ and $e = v_1 \bmod 2$. Simple minded arguments, ignoring bubbling, would give

$$D_1(\psi(\alpha^d U^p)) = c_{\alpha^d U^p} \in \mathbf{Q}.$$

To clarify the relation with the previous discussion, set

$$g_{\alpha^d U^p}(x) = \sum_e 2^{3(d+e\cdot e)/2}(-1)^{\frac{1}{2}(d+e\cdot e)}\sigma(e)(\alpha \cdot e)^d x^{\frac{1}{2}(d+e\cdot e)+1+p},$$

where now the sum runs over classes with $\frac{1}{2}(d+e\cdot e)+1+p$ a non-negative integer (and $e = v_1 \bmod 2$). Thus

$$g_{\alpha^d U^p}(x) = x f_{\alpha^d U^p}(x) + \tfrac{1}{4} c_{\alpha^d U^p}.$$

Clearly we can recover the polynomial $f_{\alpha^d U^p}$ from $g_{\alpha^d U^p}$, by removing the constant term and dividing by x, but $g_{\alpha^d U^p}$ contains a little more information. Now there is an obvious map

$$R : \underline{\underline{HF}}_*(Y) \to \mathbf{Q}[x],$$

which is not in general a homomorphism of $\mathbf{Q}[u]$-modules because of the component D_1 occurring in the definition of the product on $\underline{\underline{HF}}_*$. Clearly for $\alpha^d U^p \in A^\nu(X_1)$ we have $R(\underline{\underline{\psi}}(\alpha^d u^p) = f_{\alpha^d u^p}(x)$. The additional information we have now is summarised by

Proposition 7.15 *If the class v_1 is not zero modulo 2, and if the degree of λ in $A^\nu(X_1)$ is at most 4, then*

$$R\left(u\underline{\underline{\psi}}(\lambda)\right) = g_\lambda(x).$$

What this is asserting is that, for the restricted range of classes to which our results apply, the map $\underline{\underline{\psi}}$ is a $\mathbf{Q}[u]$-homomorphism. On the other hand, this statement encapsulates certain relations between the Floer groups and the intersection form of X_1. (Recall that the map D_1 appears as a component of the product on $\underline{\underline{HF}}$.) For example we recover the first statement of Proposition 7.3 when we take $d = 0$. Thus the relevant class λ is the identity in $A^\nu(X_1)$. If $v_1^2 = 2$, the number c_λ is just $S(X_1, v_1)$. As another example consider $\lambda = \alpha^2$, and a class v_1 with $v_1^2 = 0 \bmod 4$. Then we get a class $\underline{\underline{\psi}}(\alpha^2) \in \underline{\underline{HF}}_1$ and the map from HF_1 to $\underline{\underline{HF}}_1$ is an isomorphism, so we can lift this to a class $\psi \in HF_1$. Then Proposition 7.15 asserts that

$$D_1(\psi) = \sum \sigma(e)(\alpha \cdot e)^2, \tag{7.30}$$

where the sum runs over elements e with $e^2 = -4$ and $e = v_1 \bmod 2$. What this means, just as in the discussion following Corollary 7.4, is that if we have a 4-manifold X_1 with negative definite intersection form, and if we find classes v, α such that the sum on the right hand side of Equation 7.30 is not zero, then we can assert that the map $D_1 : HF_1 \to \mathbf{Q}$ is not zero, in particular the Floer group HF_1 of the boundary is non-trivial. Conversely, if we have a homology 3-sphere Y for which we know that $HF_1(Y) = 0$ we know that Y cannot bound any 4-manifold with this intersection form. (An exercise for the reader is to check that this sum does vanish for the standard positive definite form.) Again we shall come back to say more about these matters in Chapter 8.

7.4 Connected sums

The main goal of this Section is to prove a result, essentially due to Fukaya [26], which describes the Floer homology of a connected sum of homology 3-spheres. The result has both an algebraic, formal, aspect which will bring in the constructions of Subsection 7.3.3 and a geometric aspect. We will begin with the latter.

7.4.1 Surgery and instanton invariants

Suppose X is a closed oriented 4-manifold with $b^+(X) > 1$ and we have a moduli space M_X of instantons over X of dimension 3. Let γ be an embedded circle in X. We then have a numerical invariant $\langle \mu(\gamma), M_X \rangle$. (We will omit the discussion of orientations needed to fix signs here.) On the other hand we may perform a 'surgery' on γ to construct a new manifold X_γ. That is, we fix a trivialisation of the normal bundle of γ, so a tubular neighbourhood is identified with $S^1 \times D^3$. We cut out this neighbourhood and glue in $D^2 \times S^2$, which has the same boundary $S^1 \times S^2$, to obtain X_γ. Straightforward topology shows that X and X_γ have the same signature, while $\chi(X_\gamma) = \chi(X) + 2$. Thus there is a moduli space M_{X_γ} of instantons on X_γ of dimension $\dim M_X - 3 = 0$. This means that we have another invariant $\sharp(M_{X_\gamma})$ given by counting, with signs, the points in M_{X_γ}.

Theorem 7.16 *With suitable sign conventions,*

$$\sharp(M_{X_\gamma}) = \langle \mu(\gamma), M_X \rangle.$$

(The result we shall actually need, in Subsection 7.4.2 below, will be a little different, but for the sake of exposition we choose to begin with the case above, to avoid extraneous complications.)

The proof of Theorem 7.16 is an exercise in gluing theory. Let X_0 denote the complement of the tubular neighbourhood of γ in X, equipped with a metric with cylindrical end. Then we obtain X and X_γ respectively by gluing the manifolds $S^1 \times D^3, D^2 \times S^2$ respectively (thought of again as having cylindrical ends) to X_0, in the familiar fashion. The problem is that the cross-section of the cylinder, $S^1 \times S^2$, does not fit the standard hypotheses we have assumed up to now (the relevant bundle being in this case the trivial $SU(2)$ bundle). So, rather than appealing to any general formalism, we proceed in an *ad hoc* fashion in this special case.

The flat $SU(2)$ connections over $S^1 \times S^2$ are all reducible. They are parametrised naturally by a closed interval $[0, 1]$ with the point $t \in [0, 1]$ mapping to a connection Γ_t with holonomy

$$g_t = \begin{pmatrix} e^{i\pi t} & 0 \\ 0 & e^{-i\pi t} \end{pmatrix},$$

around the S^1 factor. We consider a small perturbation $\epsilon\eta$ of the Chern–Simons functional (where ϵ is a positive, real parameter) which has a *maximum* at the trivial connection Γ_0 and a minimum at Γ_1 and with

$$\frac{d}{dt}\eta(\Gamma_t) < 0,$$

for $t \in (0, 1)$. The deformed functional has then just two critical points Γ_0, Γ_1. Our gluing theory from Chapter 4 enables us to analyse the appropriate deformed instantons over X and X_γ in terms of the pieces in their decompositions. Straightforward index arguments tell us that the relevant moduli spaces over X_0 are

- a 3-dimensional moduli space $M^3_{X_0}$ of connections with limit Γ_1,
- a 0-dimensional moduli space $M^0_{X_0}$ of connections with limit Γ_0.

Now we turn to the other pieces in the decomposition. For the unperturbed functional, i.e. for the ordinary instanton equations, each connection Γ_t obviously extends to a unique flat, reducible, solution over $S^1 \times D^3$, whereas over $D^2 \times S^2$ only the trivial connection Γ_0 extends. After perturbation the situation is similar. Over $S^1 \times D^3$ both connections Γ_0, Γ_1 extend to reducible solutions with isotropy $SO(3)$, say Θ_0, Θ_1, while over $D^2 \times S^2$ only Γ_0 extends, again to the trivial solution which we will denote by Φ_0. Index arguments show that the obstruction spaces H^2 *vanish* for each of the connections Θ_0, Θ_1, Φ.

We now take a parallel copy γ' of the original circle γ. So we may think of γ' as being contained in X_0. Familiar arguments show that $\langle \mu(\gamma), M_X \rangle$ is the same as the invariant $\langle \mu(\gamma'), M^3_{X_0} \rangle$ computed on X_0. This is because, after deformation, the connections in the moduli space M_X are obtained from those in $M^3_{X_0}$ by gluing to the reducible solution Θ_1. Similarly the invariant $\sharp(M_{X_\gamma})$ is equal to the number of points $\sharp(M^0_{X_0})$, since the relevant connections are obtained by gluing to the reducible solution Φ. So the identity we have to prove is

$$\langle \mu(\gamma'), M^3_{X_0} \rangle = \sharp(M^0_{X_0}), \tag{7.31}$$

involving only the manifold X_0.

Lemma 7.17 *For sufficiently small ϵ there is a reducible connection B over $S^1 \times S^2 \times \mathbf{R}$ which represents a gradient line of the deformed Chern–Simons functional $\vartheta + \epsilon\eta$ running from the critical point Γ_0 to the critical point Γ_1. Moreover B is the unique connection with these properties.*

This Lemma is an instance of a general principle. Suppose one has Morse–Bott function F with a critical manifold C. If one deforms the function by adding a perturbation ϵf then for small ϵ the gradient lines of the function $F + \epsilon f$ which stay near to C are modelled by the gradient lines of the function f restricted to C. In our case the corresponding model is to take the function $\eta(\Gamma_t)$ on $[0,1]$ which plainly has a gradient line running from 0 to 1.

To outline the proof of Lemma 7.17 in more detail, observe first that it suffices to work with S^1 connections, which simplifies things somewhat. We identify the space of S^1 connections, up to a covering, with the kernel of d^* on the 1-forms, and write this vector space as a product $V_0 \times V_1$, where V_0 is the 1-dimensional space of harmonic 1-forms and V_1 is its orthogonal complement in the kernel of d^*. We write points of this space as pairs $(x, y) \in V_0 \times V_1$. The unperturbed Chern–Simons functional is given in these co-ordinates by a quadratic form in the y-variable, and its gradient is an (unbounded) linear operator $L : V_1 \to V_1$ whose spectrum is bounded away from 0. Thus the instantons on the tube, i.e. gradient lines of the unperturbed problem, are solutions $(x(t), y(t))$ of the system

$$\frac{dx}{dt} = 0, \tag{7.32}$$

$$\frac{dy}{dt} = Ly. \tag{7.33}$$

The trivial connection Γ_0 corresponds to the point $(0, 0)$ in $V_0 \times V_1$ and the connection Γ_1 to $(1, 0)$, say. Consider a perturbation $\eta(x, y)$ which is the restriction to the reducible connections of a perturbation on the $SU(2)$ connections, as considered above. The gradient of η can be expressed as a pair of components $(F(x, y), G(x, y))$ say. The fact that the perturbation extends to $SU(2)$ connections entails that F and G both vanish at the points $(0, 0)$ and $(1, 0)$. We want to solve the perturbed equations

$$\frac{dx}{dt} = \epsilon F(x, y), \tag{7.34}$$

$$\frac{dy}{dt} = Ly + \epsilon G(x, y), \tag{7.35}$$

for small ϵ. Consider the restriction of F to the line $y = 0$. As we have observed above, there is obviously a function $u(t)$ such that

$$\frac{du}{dt} = F(u(t), 0)),$$

defined for all $t \in \mathbf{R}$, with $u(t) \to 0$ as $t \to -\infty$ and $u(t) \to 1$ as $t \to +\infty$. Write

$$x(t) = u(\epsilon t) + \epsilon p(t), \tag{7.36}$$

$$y(t) = \epsilon q(t). \tag{7.37}$$

Then we need to solve the equations

$$\frac{dp}{dt} = F(u(\epsilon t) + \epsilon p, \epsilon q) - F(u(\epsilon t), 0), \tag{7.38}$$

$$\frac{dq}{dt} = Lq + \epsilon G(u(\epsilon t) + \epsilon p, \epsilon q). \tag{7.39}$$

We can write these equations schematically as $\mathcal{F}_\epsilon(p, q) = 0$. The approach is to find a solution using the implicit function theorem. The relevant linearised operator $\mathcal{L}$ is given by

$$\mathcal{L}(p, q) = \left(\frac{dp}{dt}, \frac{dq}{dt} - Lq\right),$$

and this is an operator of the type we have studied in Chapter 3 so can be inverted on Sobolev spaces using a weight function which behaves as $e^{\alpha|t|}$ as $t \to \pm\infty$, for small positive α. On the other hand, if we start with the approximate solution (p_0, q_0) given by $p_0 = 0, q_0 = 0$ we have

$$\mathcal{F}_\epsilon(p_0, q_0) = (0, \epsilon G(u(\epsilon t), 0)).$$

We assume that the critical points of the perturbation are non-degenerate; this means that $u(t)$ converges exponentially as $t \to \pm\infty$ and hence, since G vanishes at $(0, 0)$ and $(1, 0)$, the term $G(u(\epsilon t, 0)$ decays exponentially as $t \to \pm\infty$. So if α is sufficiently small the initial error $\mathcal{F}_\epsilon(p_0, q_0)$ lies in our weighted Sobolev space, and has small norm when ϵ is small. It is then a straightforward exercise, similar to the gluing theorems for instantons in Chapter 4, to show that there is a solution to the problem close to (p_0, q_0). The discussion of uniqueness is similar.

With this Lemma in place, we may now analyse the end of the moduli space $M^3_{X_0}$ in the usual way. For each point in $M^0_{X_0}$ we get a component of the end, modelled on $S^2 \times (0, \infty)$, by gluing the corresponding connection to the gradient line from Γ_0 to Γ_1. The 2-sphere factor arises from the gluing parameter $S0(3)$ modulo the isotropy S^1 of the gradient line. The formula expresses the fact that the cohomology class $\mu(\gamma')$ can be localised on these ends. To do this we fix a specific family of representatives for the cohomology class. This falls into the same line of ideas as the proof of Proposition 7.12. Fix a base point x_0 in X_0. For large L we consider a loop σ_L in X_0 which begins at x_0, travels a distance L down the cylindrical end of X_0, then runs once around the S^1 factor in the cross-section before retracing its route back to x_0. Now, as in Subsection 7.3.2, we assume that the base point fibration over $M^3_{X_0}$, associated to the base point x_0, lifts to an $SU(2)$ bundle and fix a trivialisation of this bundle. So the holonomy around σ_L gives a map

$$h_L : M^3_{X_0} \to SU(2).$$

Now fix $\tau \in (0,1)$ and a representative for the generator of $H^3(SU(2)) = H^3(S^3)$ supported in a small neighbourhood of the point $g_\tau \in SU(2)$. Pulling this back to the moduli space by h_L we obtain the desired family of representative ϕ_L of $\mu(\gamma')$, one for each large L.

We now claim that as $L \to \infty$ the support of ϕ_L moves into the ends of $M^3_{X_0}$. Indeed for any fixed point $[A] \in M^3_{X_0}$ the trace of the holonomy around the S^1 factor tends to -2 at infinity, since the limit is $\Gamma_1 = -1$ whereas in the support of ϕ_L the trace must be close to $2\cos(\pi\tau) > -2$. So the proof is completed by showing that the contribution of each end is 1. Consider the description of an end as $S^2 \times (0,\infty)$, where the S^2 arises from the gluing parameter and $(0,\infty)$ from translation of the gradient line B. In this description the map h_L is given approximately by

$$(z,s) \mapsto \exp(b(z)c(s-L)),$$

where we regard S^2 as the sphere in the Lie algebra of $SU(2)$, b is a rotation of the sphere and $c : (0,\infty) \to (0,\pi)$ is a diffeomorphism. It is clear then that the restriction of ϕ_L is supported in a neighbourhood of the point $(z_0, c^{-1}(\pi\tau) + L)$ in the end, and evaluates to 1.

7.4.2 The $\mathrm{Hom}_{\mathcal{F}}$-complex and connected sums

We now proceed to the formal, algebraic, part of the discussion. Recall that in Subsection 7.3.3 we defined, for any homology cobordism W between homology 3-spheres Y, Y', a particular chain map from $\tilde{C}_*(Y)$ to $\tilde{C}_*(Y')$ lying in $\mathrm{Hom}_{\mathcal{F}}(\widetilde{CF}_*, \widetilde{CF}'_*) \oplus \mathbf{Q}$, and this was independent of choices up to a chain homotopy in $\mathrm{Hom}_{\mathcal{F}} \oplus \mathbf{Q}$. We saw that the homomorphisms $\mathrm{Hom}(\widetilde{CF}_*, \widetilde{CF}'_*)$ can be made into an $(\mathcal{F},\sigma)$-complex via a filtration

$$\mathrm{Hom}_{\mathcal{F}}(Y,Y') \subset \mathrm{Hom}_{\mathcal{F}}(Y,Y') \oplus \mathbf{Q} \subset \mathrm{Hom}(\widetilde{CF}_*, \widetilde{CF}'_*).$$

Here, and in what follows, we just write $\mathrm{Hom}_{\mathcal{F}}(Y,Y')$ for $\mathrm{Hom}_{\mathcal{F}}(CF_*(Y), CF_*(Y'))$

Theorem 7.18 *The $(\mathcal{F},\sigma)$-complex associated to a connected sum $\overline{Y}\sharp Y'$ is equivalent in $\mathcal{C}$ to $\mathrm{Hom}(\widetilde{CF}_*(Y), \widetilde{CF}_*(Y'))$. In particular,*

(1) *the Floer homology $HF_*(\overline{Y}\sharp Y')$ can be computed as the homology of the complex $\mathrm{Hom}_{\mathcal{F}}(Y,Y')$,*
(2) *the homology $\overline{HF}(\overline{Y}\sharp Y')$ can be computed as the homology of the complex $\mathrm{Hom}_{\mathcal{F}}(Y,Y') \oplus \mathbf{Q}$.*

The change of orientation on one factor may seem strange and one can easily avoid it by discussing a tensor product in the category $\mathcal{C}$; however, we find the Hom formulation slightly more convenient.

The complex $\mathrm{Hom}_{\mathcal{F}}(Y, Y')$ is itself a filtered complex, with a filtration given by the terms

- $\mathrm{Hom}(\mathbf{Q}, C'_*)$,
- $\mathrm{Hom}(\sigma C_*, C'_*) \oplus \mathrm{Hom}(\mathbf{Q}, C'_*)$,
- $\mathrm{Hom}(C_*, C'_*) \oplus \mathrm{Hom}(\mathbf{Q}, C'_*) \oplus \mathrm{Hom}(\sigma C_*, \mathbf{Q})$,
- $\mathrm{Hom}_{\mathcal{F}}(Y, Y')$.

Here the first term corresponds to the maps $\underline{\lambda}$ with $\lambda_1 = \lambda_2 = \lambda_3 = 0$, the second term to maps with $\lambda_1 = \lambda_3 = 0$ and the third term to maps with $\lambda_1 = 0$.

This filtration implies that there is a spectral sequence converging to the Floer homology of $\overline{Y} \sharp Y'$ whose E_1 term is (ignoring the grading) given by the sum of two copies of $\mathrm{Hom}(HF_*(Y), HF_*(Y'))$, $\mathrm{Hom}(HF_*(Y), \mathbf{Q})$ and $HF_*(Y')$. The differentials in the spectral sequence are various maps induced by $D_1, D_2, U, D'_1, D'_2, U'$.

Before beginning the proof of Theorem 7.18 it may be worthwhile to explain why the result should not be surprising. Suppose for simplicity that the flat connections on Y, Y' are non-degenerate, so we have finite sets $\{\rho_\alpha\}, \{\sigma_\beta\}$ of irreducible flat connections over Y, Y' respectively. The irreducible flat connections over the connected sum are of three kinds:

- connections obtained by gluing ρ_α to σ_β, for each pair (α, β); for each such pair we get a family of flat connections over the connected sum parametrised by a copy of $SO(3)$;
- connections obtained by gluing ρ_α over Y to the trivial connection over Y';
- connections obtained by gluing a σ_β to the trivial connection over Y.

In the second and third cases the connections that are constructed are isolated. Now if we perturb the Chern–Simons functional over the connected sum the $SO(3)$ families of the first type break up into isolated connections, and we know from Morse–Bott theory that in rational cohomology each pair (α, β) should contribute two generators to the resulting description of the Floer homology of the connected sum. So at the level of chain *groups* this straightforward analysis gives precisely the picture we would expect from Theorem 7.18. The point of the theorem is that it shows that the Floer differentials are compatible with this analysis of the flat connections, which is not at all easy to prove by a direct attack. (But see [32].)

For simplicity we shall just prove the first item in Theorem 7.18, that the Floer homology of the connected sum can be computed from the complex $\mathrm{Hom}_{\mathcal{F}}$, leaving the proof of the general statement for the interested reader. Our task then is to define chain maps

$$\alpha : C_*(\overline{Y}\sharp Y') \to \mathrm{Hom}_{\mathcal{F}}(Y, Y'), \quad \beta : \mathrm{Hom}_{\mathcal{F}}(Y, Y')' \to C_*(\overline{Y}\sharp Y'),$$

and then to prove that the two composites give the identity on homology. As usual these maps will be induced by appropriate cobordisms.

Consider in general a cobordism V from a disjoint union of Y and another homology 3-sphere P to Y'. Thus V has three oriented boundary components $\overline{Y}, \overline{P}, Y'$. In just the same way as we defined a map for a cobordism W in Subsection 7.3.3 we can define a chain map

$$\alpha_V : C_*(P) \to \mathrm{Hom}_{\mathcal{F}}(Y, Y').$$

This records data from counting points in 0-dimensional moduli spaces over V and from 3-dimensional moduli spaces over V on which we evaluate the holonomy along a path from Y to Y'. The only point to notice is that we map into $\mathrm{Hom}_{\mathcal{F}}$ rather than $\mathrm{Hom}_{\mathcal{F}} \oplus \mathbf{Q}$, since we do not encounter the trivial connection over V (because we are only considering irreducible connections over P). Likewise, if U is a cobordism from Y' to the disjoint union of Y and another manifold Q, we get a chain map

$$\widehat{\beta_U} : C_*(\overline{Q}) \to \mathrm{Hom}_{\mathcal{F}}(Y', Y).$$

Now recall from Subsection 7.3.3 that a map $\underline{\lambda}$ in $\mathrm{Hom}_{\mathcal{F}}(Y, Y')$ is determined by four components $\lambda_1, \ldots, \lambda_4$ which – suppressing the involutions σ, σ' – can be taken to be maps

$$\lambda_1, \lambda_2 : \widetilde{CF}(Y) \to \widetilde{CF}(Y'),\ \lambda_3 : CF_*(Y) \to \mathbf{Q},\ \lambda_4 : \mathbf{Q} \to CF_*(Y').$$

A map μ in $\mathrm{Hom}_{\mathcal{F}}(Y', Y)$ is determined by four components μ_i, interchanging Y and Y'. We define a bilinear pairing by

$$B(\underline{\lambda}, \underline{\mu}) = \mathrm{Tr}(\mu_1\lambda_2) + \mathrm{Tr}(\mu_2\lambda_1) + \mathrm{Tr}(\mu_3\lambda_4) + \mathrm{Tr}(\mu_4\lambda_3).$$

Lemma 7.19 *The pairing B defines an isomorphism between* $\mathrm{Hom}_{\mathcal{F}}(Y', Y)$ *and the dual complex* $\mathrm{Hom}_{\mathcal{F}}(Y, Y')^*$.

We leave the proof to the reader. Note that the pairing is *not* induced by taking the trace of the composite map from $\widetilde{CF}(Y)$ to $\widetilde{CF}(Y)$.

Using Lemma 7.19, and the familiar duality on the ordinary Floer chains, we can use $\widehat{\beta_U}$ to get a chain map

$$\beta_U : \mathrm{Hom}_{\mathcal{F}}(Y, Y') \to C_*(Q).$$

Our first task is to give a geometric interpretation to the composite map

$$\beta_U \alpha_V : C_*(P) \to C_*(Q).$$

To do this we glue U and V along their two common boundary components, so we get a 4-manifold Z which is a cobordism from P to Q. Paths in U, V joining Y and Y' glue together to give a loop in Z.

Proposition 7.20 *The composite $\beta_U \alpha_V$ is chain-homotopic to the map $\psi(\gamma) \in C_*(P)^* \otimes C_*(Q)$ defined by evaluating the class $\mu(\gamma)$ on 3-dimensional moduli spaces of instantons over Z, with irreducible limits over P, Q.*

The proof follows familiar lines. Imagine constructing Z by first gluing U to V along Y' giving a manifold Z^* say, with four boundary components: P, Q and two copies of Y. We have a path γ' in Z^* which runs from one copy of Y to the other. Now if λ_i and μ_j are the maps defined by instanton moduli spaces over U, V in the manner of Subsection 7.3.3, the first three terms $\mu_1\lambda_2 + \mu_2\lambda_1 + \mu_3\lambda_4$ in the pairing $B(\underline{\lambda}, \underline{\mu})$ correspond geometrically to evaluating the holonomy along γ' on 3-dimensional moduli spaces of instantons over Z^*. The first two terms correspond to instantons obtained by gluing 3-dimensional and 0-dimensional moduli spaces over U and V with an irreducible limit over Y'. The third term corresponds to gluing instantons in 0-dimensional moduli spaces with the trivial limit over Y' and the holonomy detects the gluing parameter, just as in Proposition 7.8. If we now proceed to construct Z from Z^* by gluing the two copies of Y we can apply the same approach to analyse the pairing of $\mu(\gamma)$ with a 3-dimensional moduli space over Z. This pairing is the sum of two pieces. The first is obtained by evaluating the holonomy along γ^* on 3-dimensional moduli space of instantons over Z^*, for connections with the same irreducible limits over the two copies of Y. The contribution from these corresponds to the three terms $\mu_1\lambda_2 + \mu_2\lambda_1 + \mu_3\lambda_4$, as above. The second piece comes from gluing connections in 0-dimensional moduli spaces over Z^* with the trivial limit over the two copies of Y, and the holonomy around γ detects the resulting gluing parameter. The contribution from this corresponds to the fourth term $\mu_4\lambda_3$ in the formula for B.

Now let Z_γ be the manifold obtained from Z by performing a surgery on the loop γ, as in Subsection 7.4.1. So Z_γ is another cobordism from P to Q.

Corollary 7.21 *The map $\beta_U\alpha_V : C_*(P) \to C_*(Q)$ is chain-homotopic to the ordinary map on the Floer groups induced by the cobordism Z_γ.*

This follows from the previous result by the same argument as used in the proof of Theorem 7.16 – the boundary components play no real role in the argument.

Now suppose that the manifold Q can be identified with $\overline{P}$. Thus we have

$$\alpha_V : CF_*(P) \to \mathrm{Hom}_{\mathcal{F}}(Y, Y'), \quad \beta_U : \mathrm{Hom}_{\mathcal{F}}(Y, Y') \to HF_*(P).$$

To get a geometric interpretation of the composite

$$\alpha_V\beta_U : \mathrm{Hom}_{\mathcal{F}}(Y, Y') \to \mathrm{Hom}_{\mathcal{F}}(Y, Y'),$$

we glue U and V along the two copies of P in their boundaries to get a manifold W with four boundary components: two copies of Y' and two copies of $\overline{Y}$. Working through the definitions and using familiar gluing arguments we find that the various components of the chain map $\alpha_V\beta_U$ are as follows.

- The component from λ_1 to λ_2 is given by counting points in 0-dimensional moduli spaces over W with four irreducible limits.
- The component from λ_2 to λ_1 is given by evaluating the product of cochains defined by two holonomy maps on 6-dimensional moduli spaces with four irreducible limits.
- The components from λ_1 to λ_1 and from λ_2 to λ_2 are given by evaluating the two different holonomy maps on 3-dimensional moduli spaces with four irreducible limits.

- The components from λ_2 to λ_3 and λ_4, and from λ_3 and λ_4 to λ_1, are given by evaluating holonomy maps on 3-dimensional moduli spaces with one trivial limit and three irreducible limits.
- The components from λ_1 to λ_3 and λ_4, and from λ_3 and λ_4 to λ_2, are given by counting points in 0-dimensional moduli spaces with one trivial limit and three irreducible limits.
- The components from λ_3 to λ_3 and from λ_4 to λ_4 are given by counting points in 0-dimensional moduli spaces with two trivial limits and two irreducible limits, where the trivial limits are over one of the pairs $\overline{Y}, Y'$ in the boundary.
- The components from λ_3 to λ_4 and from λ_4 to λ_3 are given by counting points in 0-dimensional moduli spaces with two trivial limits and two irreducible limits, where the trivial limits are over either the two copies of $\overline{Y}$ or the two copies of Y' in the boundary.

We can now proceed to complete the proof. There is a standard cobordism between the connected sum of two manifolds and their disjoint union. We take this as our manifold V, with suitable orientations, and let U be the same manifold with orientation reversed. Thus P is the connected sum $\overline{Y}\sharp Y'$. It is then clear that the surgeried manifold Z_γ is the product $(\overline{Y}\sharp Y') \times [0,1]$ so by Corollary 7.21 the composite $\beta_U \alpha_V$ induces the identity map on $HF_*(\overline{Y}\sharp Y')$. On the other hand the manifold W can likewise be identified as the internal connected sum of $\overline{Y} \times [0,1]$ and $Y' \times [0,1]$. So $\alpha_V \beta_U$ is chain-homotopic to a map $R : \mathrm{Hom}_{\mathcal{F}} \to \mathrm{Hom}_{\mathcal{F}}$ say, where R is defined by the same procedure as above but using instanton moduli spaces for a metric on W in which the neck of the connected sum is made very long. These instantons over W are given by instantons over the two tubes, glued over the neck. If the connections over the tubes are irreducible there is an $SO(3)$ gluing parameter. It follows then that all the terms above vanish except for

- those from λ_i to λ_i for $i = 1, 2, 3, 4$,
- the component from λ_2 to λ_1, given by 6-dimensional moduli spaces.

The maps in the first case just give the identity maps on the four components. The only difficulty comes from the second case: the component from λ_2 to λ_1, at this chain level, might not vanish. However, we can get around this without further work. It follows from the above analysis that the chain map R is an isomorphism from $\mathrm{Hom}_{\mathcal{F}}$ to itself (since it differs from the identity by an off-diagonal term), so certainly induces an isomorphism on homology. It follows then that $\alpha_V \beta_U$ induces an isomorphism on homology. But since we know, from the first part of the proof, that $\beta_U \alpha_V$ induces the identity map on $HF_*(\overline{Y}\sharp Y')$, it follows that $\alpha_V \beta_U$ must induce the identity on the homology of $\mathrm{Hom}_{\mathcal{F}}$, and the proof is complete.

8

Further directions

In this last Chapter we discuss some issues which lie beyond the core of this book, and which offer scope for further work.

8.1 Floer homology for other 3-manifolds

In this book we have defined and studied Floer groups in two cases: either for $SU(2)$ bundles over homology 3-spheres or for admissible $U(2)$ bundles. A natural question is whether one can define Floer groups more generally. Of course the whole problem revolves around the treatment of the reducible connections, and the discussion of Chapter 7 suggests that one should not expect there to be a single answer to this question – as opposed to a number of variants of Floer's basic construction.

Suppose then that Y is any compact oriented 3-manifold and we consider connections on an $SU(2)$ bundle over Y. (The discussion applies equally well to the $U(2)$ case, and indeed in large part goes over to general gauge groups.) We may perturb the Chern–Simons functional with a perturbation η to arrange that all critical points are non-degenerate; in particular there are a finite number of irreducible critical points. The simplest thing one can do is to set up a chain complex (CF^{η}_*, ∂) using just these irreducibles. The whole theory of Chapters 5 and 6 extends without essential change to this case. In particular the same reasoning shows that $\partial^2 = 0$ because we do not encounter reducible critical points in the factorisations corresponding to the boundary of 1-dimensional moduli spaces. Thus we obtain some Floer homology groups $HF^{\eta}_*(Y)$ in this general situation. Moreover the solutions of suitably perturbed instanton equations yield invariants for 4-manifolds-with-boundary Y, satisfying the familiar gluing rules. This simple approach is not really satisfactory however because the Floer homology defined by the complex CF_*^η will in general depend on the perturbation η chosen. As usual, the same phenomenon can be seen in finite-dimensional analogues. Suppose for example we consider the complex plane $\mathbf{C}$ (as an analogue of the space of connections $\mathcal{A}$), with the standard action of the circle (as an analogue of the gauge group). The circle-invariant function $\Phi(z) = |z|^4$ has a single degenerate critical point at $z = 0$, which is the fixed point of the action. We consider the family of perturbations

$$\Phi_\epsilon(z) = |z|^4 + \epsilon|z|^2.$$

For non-zero ϵ the function Φ_ϵ is non-degenerate, regarded as a circle-invariant function. When ϵ is positive the only critical point is at the origin whereas if ϵ is negative there is another orbit of critical points $|z| = \sqrt{\frac{\epsilon}{2}}$. Down on the quotient space $\mathbf{C} \setminus \{0\}/S^1 \cong \mathbf{R}^+$ – analogous to $\mathcal{B}^*$ – there is a single critical point for $\epsilon < 0$ which is not present when $\epsilon > 0$. Clearly the problem is that as the parameter ϵ increases through zero the 'irreducible' critical point is sucked into the reducible one and disappears. It is easy to reproduce this phenomenon any time one has a fixed point of a circle action on a manifold – one can change the functions Φ_ϵ slightly to make them equal to Φ for large $|z|$ and then glue this model into a neighbourhood of the fixed point. In such a case the homology groups, constructed using the analogue of the CF_*^η complex, must be different for positive and negative values of ϵ. This follows from the fact that just one critical point is lost so the Euler characteristics in the two cases must differ by 1. In turn exactly the same thing can occur in the infinite-dimensional Floer theory situation. Going back to the case of a homology 3-sphere, one sees that the essential thing there is that the only reducible critical point is the trivial connection, which is already non-degenerate for the original Chern–Simons functional. The key point is that in Chapter 5 we chose our perturbations to be sufficiently small so that the Hessian at θ did not acquire a kernel, which meant that we did not encounter this phenomenon.

The machinery of equivariant homology and cohomology overcomes this defect in the finite-dimensional situation. Going back to our model example on $\mathbf{C}$ we observe that when $\epsilon < 0$ there is a gradient line running from the 'reducible' critical point $z = 0$ to the irreducible critical point. Suppose we have a compact manifold M with a circle action having a single fixed point O, and we consider a family of invariant functions Φ_ϵ on M which behave like this model near the critical point. Then for any non-zero ϵ we can form complexes $\overline{\overline{C}}^\epsilon_*, \underline{\underline{C}}^\epsilon_*$ in the same manner as our complex $\overline{\overline{CF}}_*, \underline{\underline{CF}}_*$. That is, we take a generator for each irreducible critical point to get C^ϵ_* and adjoin copies of $\mathbf{Q}[[y]], \mathbf{Q}[x]$ respectively, corresponding to the fixed point. The formal variables x, y in this case have degree 2, corresponding to the generator of $H^*(\mathbf{CP}^\infty)$. The differentials in $\overline{\overline{C}}^\epsilon_*, \underline{\underline{C}}^\epsilon_*$ are defined just as in Subsection 7.3.4. In $\overline{\overline{C}}$ we take

$$\partial + yD_1 \circ h + \tfrac{1}{2}y^2 D_1 \circ h^2 + \cdots,$$

and in $\underline{\underline{C}}$ we take

$$\underline{\underline{\partial}}(\psi, f(x)) = \partial\psi + f(h)D_2(1).$$

Here D_1 is defined by counting gradient lines from irreducible critical points to the fixed point, and D_2 by counting those from the fixed point to the irreducibles, in just the same fashion as our complexes of Chapter 7. The cap product map $h : C^\epsilon_* \to C^\epsilon_{*-2}$ is defined using a representative for the first Chern class of the S^1 fibration $M \setminus \{O\} \to M \setminus \{O\}/S^1$, in the manner of Subsection 7.3.2. Now the homology of these complexes does not change as ϵ crosses 0 – more precisely there is a canonical isomorphism between the homology groups for different values of ϵ. Of course, following the discussion of Subsection 7.3.4 above, and [5], the groups we get are the standard equivariant homology and cohomology groups of the manifold M with the circle action. However, it is instructive to see directly what happens as ϵ increases through 0. We write $\underline{\underline{H}}^\pm_*$ for the $\underline{\underline{H}}^\epsilon_*$ when ϵ is close to zero and positive or negative respectively; similarly for $\overline{\overline{H}}^\pm_*$. First consider the case of $\underline{\underline{C}}$. When $\epsilon < 0$ there is a generator $\langle p \rangle$ say, corresponding to the 'irreducible' critical point. So we have

$$\mathbf{Q}\langle p \rangle \oplus \mathbf{Q}[x] \subset \underline{\underline{C}}^\epsilon_*.$$

The gradient line running from O to p means that the map D_2 sends $1 \in \mathbf{Q}[x]$ to $\langle p \rangle$. Let us suppose that there are no other relevant gradient lines, involving other critical points. Then in forming the homology we 'lose' two generators, so we are left with a copy of $x\mathbf{Q}[x] \subset \underline{\underline{H}}^-_*$. When ϵ is positive we have a local contribution of just $\mathbf{Q}[x]$ to $\underline{\underline{C}}^\epsilon_*$ and no differential so we get a copy of $\mathbf{Q}[x] \subset \underline{\underline{H}}^+_*$. The canonical isomorphism between $\underline{\underline{H}}^-_*$ and $\underline{\underline{H}}^+_*$ maps the powers $x^n \in \underline{\underline{H}}^-_*$ to $x^{n-1} \in \underline{\underline{H}}^+_*$. The point is that when ϵ passes through zero the *index* of the fixed point O changes by 2, which accounts for the shift in degree. In the case of $\overline{\overline{C}}_*$ the relevant differential is D_1 and this is zero in both cases (since there is no gradient line *from p to O*). Thus we have a local contribution to $\overline{\overline{H}}_*^-$ of

$$\langle p \rangle \oplus \mathbf{Q}[y]$$

and to $\overline{\overline{H}}_*^+$ of $\mathbf{Q}[y]$. The isomorphism between these two maps $\langle p \rangle \in \overline{\overline{H}}_*^-$ to $1 \in \mathbf{Q}[y] \subset \overline{\overline{H}}_*^+$ and $y^n \in \mathbf{Q}[y] \subset \overline{\overline{H}}_*^-$ to $y^{n+1} \in \overline{\overline{H}}_*^+$. Again, the shift in degree is a reflection of the change of index of O.

We may take all this discussion over to the gauge theory case. For a fixed perturbation of the Chern–Simons functional we have the chain complex CF_*^η, ∂ defined by the irreducible critical points. Let us discuss the case of $\underline{\underline{CF}}_*$. For each reducible critical point σ, with stabiliser $\Gamma_\sigma \subset SO(3)$, we take a copy of the rational cohomology group $H^*(B\Gamma_\sigma)$. The most important cases are when Γ_σ is the circle or $SO(3)$ and the cohomology rings are polynomial rings with generators in dimension 2 or 4 respectively. The direct sum of these, over all the reducibles (and with suitable grading), gives the contribution CF_*^{red} from reducibles in the desired chain complex and we set

$$\underline{\underline{CF}}_* = CF_*^\eta \oplus CF_*^{\text{red}}.$$

Now we want to discuss the differential. The component from CF_*^η to CF_*^η is just ∂. Let us write $\underline{D}_2$ for the component of the differential mapping CF_*^{red} to CF_*^η. Then $\underline{D}_2$ is a sum of pieces $\underline{D}_2 = \bigoplus_p D_2^p$. The first piece D_2^0 is given by the map

$$D_2^0 : \bigoplus_\sigma H^0(B\Gamma_\sigma) = \bigoplus_\sigma \mathbf{Q} \to \mathbf{CF}_*^\eta,$$

which is defined just as for the map D_2 in Chapter 5, counting instantons with one reducible limit at ∞. To define the other pieces, we have to proceed a little differently from Subsection 7.3.4, since when $\Gamma_\sigma = S^1$ the generator of $H^2(BS^1)$ does not correspond to a cohomology class over the space $\mathcal{B}_Y^*$ as was the case for $SO(3)$. We consider a $(p+1)$-dimensional moduli space $M_{\sigma\rho}$ of instantons with one reducible limit σ at $-\infty$ and one irreducible limit ρ at $+\infty$. There is a Γ_σ bundle over $M_{\sigma\rho}$ defined by 'taking a base point at $-\infty$'. We choose suitable representatives for the characteristic classes of Γ_σ bundles, associated with $H^p(B\Gamma_\sigma)$, then evaluate these representatives on the reduced space $M_{\sigma\rho}/\mathbf{R}$. This gives a map

$$n_{\sigma\rho} : H^p(B\Gamma_\sigma) \to \mathbf{Q} = \mathbf{Q}\langle \rho \rangle.$$

Taking the sum of these over all σ, ρ gives D_2^p.

We now come to an important point. Suppose that *there are no non-trivial reducible instantons on the tube $Y \times \mathbf{R}$, or more precisely that there are no non-trivial reducible solutions of the relevant perturbed equation.*

If this hypothesis holds the sketch above outlines the complete story: there are no further components of the differential in $\underline{\underline{CF}}_*$. Of course the outline needs to be filled in, but there are no essential new difficulties in showing that this procedure defines a chain complex, with homology groups $\underline{\underline{HF}}_*(Y)$ that are independent of the small perturbations used. This is essentially the theory developed in [6]. In a similar fashion we can define groups $\overline{\overline{HF}}_*(Y)$.

When the hypothesis above does not hold new difficulties appear. Consider, for example, a pair of reducible critical points σ_1, σ_2 with stabilisers $\Gamma_{\sigma_i} \cong S^1$, and a reducible instanton a from σ_1 to σ_2. There are two different indices we can assign to a, first the index $\mathrm{ind}_{SU(2)}(a)$ which gives the formal dimension of the moduli space of $SU(2)$ instantons containing a, and second the index $\mathrm{ind}_{S^1}(a)$ which gives the formal dimension of the moduli space of S^1 instantons. The difference of these indices is the index of a complex linear operator. Suppose for example that $\mathrm{ind}_{S^1}(a) = 0$ and $\mathrm{ind}_{S^1}(a) = 2q > 0$. Then we can suppose that a is an isolated point in the moduli space of S^1 instantons from σ_1 to σ_2 and there is an 'obstruction space' $H^2_a \cong \mathbf{C}^q$. Now suppose we have an irreducible instanton A from σ_2 to an irreducible critical point ρ and we consider the gluing problem of gluing A to a to construct an instanton from σ_1 to ρ. The presence of the obstruction space H^2_a means that we cannot normally do this. Instead we get a section of a $\mathbf{C}^q$ bundle over $M'_{\sigma_2\rho}$ and the gluing problem can be solved at the zeros of this section (assuming these are transverse). In other words, to find the contribution to the end of $M_{\sigma_1\rho}$ arising from factorisations through σ_2 we need to find the zeros of this section. What this means is that $\partial \circ \underline{D}_2$ need not vanish, and to get a chain complex we need to add in data arising from these obstruction bundles and sections. The general picture we expect is that there is now another component of the differential in $\underline{\underline{CF}}_*$, say

$$E : CF^{\mathrm{red}}_* \to CF^{\mathrm{red}}_*.$$

In the case above the contribution to E arising from a will be the map

$$H^*(B\Gamma_{\sigma_1}) = H^*(\mathbf{CP}^\infty) \to H^*(B\Gamma_{\sigma_2}) = H^*(\mathbf{CP}^\infty),$$

which maps the standard generator h^r of $H^{2r}(\mathbf{CP}^\infty)$ to h^{r-q} if $r \geq q$ and to 0 if $r < q$.

As usual, of course, there is no difficulty in mimicking this whole discussion in a finite-dimensional context. However, in the Floer theory case great care needs to be taken on account of bubbling phenomena. The point is that the analysis of the ends of moduli spaces in the presence of reducible instantons on the tube is close to the analogous problem over compact manifolds when $b^+ = 0$, which is fraught with difficulties as we have explained in Subsection 7.3.5 (and which we return to below). Thus while the discussion above certainly gives part of the story – the analogue of the 'naive' theory in Subsection 7.3.5 – it is not clear to the author whether this needs to be modified to take account of bubbling, and if so precisely what additional terms are required. Certainly in many specific cases one can arrange that there is no contribution from these reducible instantons, and so obtain a definition of the equivariant Floer groups, but more work is required for a complete theory.

One of the drawbacks of the equivariant theory $\underline{\underline{HF}}_*, \overline{\overline{HF}}_*$ – even in cases when the technical difficulties discussed above have been overcome – is that the groups are necessarily infinite-dimensional. On the one hand this makes them less useful as a source of potential 3-manifold invariants: for example one cannot take the Euler characteristic. On the other hand the infinite-dimensionality is unnecessary in many cases in so far as the gluing problem is concerned. If we have 4-manifolds X_1, X_2 with boundary $Y, \overline{Y}$ respectively and if there are no reducible instantons over X_1 and X_2 then the invariants of $X = X_1 \cup_Y X_2$ can be described in terms of relative invariants with values in the finite-dimensional groups $HF_*^\eta(Y)$ defined by some fixed perturbation η. The extra refinement of passing to $\underline{\underline{HF}}, \overline{\overline{HF}}$ involves redundant information from this point of view. Thus we would like to have some *finite-dimensional* Floer groups which on the one hand are independent of perturbations, but on the other hand serve as repositories for the relative invariants in this situation. One can hope to do this as follows. Assume that one has defined a satisfactory map E above, and hence a differential on $\underline{\underline{CF}}_*(Y)$. From the form of the differential the natural inclusion gives a chain map $CF_*^\eta \to \underline{\underline{CF}}_*$ and hence an induced map on homology. Similarly there is a map from $\overline{\overline{HF}}_*(Y)$ to $HF_*^\eta(Y)$. Composing these we get a map

$$\pi : \overline{\overline{HF}}_*(Y) \to \underline{\underline{HF}}_*(Y).$$

If the equivariant theory has the expected properties this map will be a topological invariant. In particular we may define

$$\widehat{HF}_*(Y) = \operatorname{Im} \pi \subset \underline{\underline{HF}}_*(Y).$$

This group will be finite-dimensional and one can define relative invariants for 4-manifolds with boundary, provided there are no reducible instantons over the 4-manifolds, with values in $\widehat{HF}$, essentially because π factors through HF_*^η. For example in the case of homology 3-spheres $\widehat{HF}_*(Y)$ is obtained from $HF_*(Y)$ by taking

$$\begin{aligned}
\widehat{HF}_i &= HF_i \text{ if } \ i \neq 0,1,4,5, \\
\widehat{HF}_5 &= \bigcap_j \ker D_1 u^{2j+1} \subset HF_5, \\
\widehat{HF}_1 &= \bigcap_j \ker D_1 u^{2j} \subset HF_1, \\
\widehat{HF}_0 &= HF_0 \Big/ \left(\bigoplus \operatorname{Im} u^{2j+1} D_2 \right), \\
\widehat{HF}_4 &= HF_4 \Big/ \left(\bigoplus \operatorname{Im} u^{2j} D_2 \right).
\end{aligned}$$

These groups have been introduced and studied by Froyshov [25].

8.2 The blow-up formula

The gluing theory we developed in Chapter 7 is incomplete, as we explained in Subsection 7.3.4, because it does not take into account the additional boundary of moduli spaces arising from the 'bubbling' phenomenon. The prototype here is the case when the manifold Y is a 3-sphere, so the gluing problem involves a connected sum. In particular we consider the case of a 4-manifold $X = X_1 \sharp X_2$ where X_2 is a connected sum of copies of $\overline{\mathbf{CP}}^2$. (We will work with $SU(2)$ connections throughout this Section, although the whole discussion goes over to the $U(2)$ case.) The gluing problem asks how we can calculate pairings $\langle \mu(\alpha)^d, M_X \rangle$, for $\alpha \in H_2(X_2)$, in terms of invariants for the manifold X_1. Plainly it suffices to consider the case when $X_2 = \overline{\mathbf{CP}}^2$ and α is the standard generator E of $H_2(\overline{\mathbf{CP}}^2)$. As we have seen in Chapters 6 and 7, this is easy when d is small since the bubbling phenomena does not then occur, for dimensional reasons. For some larger values of d, direct calculations were made by Orszvath [38] and Leness [31]. These calculations involve a very careful description of the boundary of the moduli spaces, and become more and more complicated as d increases. In a remarkable paper [20], Fintushel and Stern showed how a comparatively simple but indirect argument gives the complete answer to this problem. The strategy is to consider a double connected sum $X_1 \sharp \overline{\mathbf{CP}}^2 \sharp \overline{\mathbf{CP}}^2$, so there are two homology classes E_1, E_2 in the two factors, represented by disjoint embedded spheres of self-intersection -1. Joining these spheres by a tube we get an embedded surface Σ of self-intersection -2. The boundary of a tubular neighbourhood of Σ is a copy of $\mathbf{RP}^3$, so we have another decomposition

$$X_1 \sharp \overline{\mathbf{CP}}^2 \sharp \overline{\mathbf{CP}}^2 = X_0 \sharp_{\mathbf{RP}^3} Z$$

say. While the 3-manifold $\mathbf{RP}^3$ does not precisely fit into the framework of the gluing theory we developed in the earlier Chapters, since it is a rational but not integral homology sphere, it is straightforward to extend the familiar techniques to this case. In particular there are straightforward formulae expressing invariants defined by pairings $\langle \mu(\Sigma)^r \mu(\beta)^s, M\rangle$, with $\beta \in H_2(X_0)$, in terms of moduli spaces over X_0, provided that r is small; in fact $r \leq 4$. Now the homology class of Σ is $E_1 + E_2$. The key observation is that $\beta = E_1 - E_2$ is contained in $H_2(X_0)$ (since $\beta \cdot \Sigma = 0$). This means that we have a good understanding of pairings

$$\langle \mu(E_1 + E_2)^4 \mu(E_1 - E_2)^s, M\rangle,$$

for any value of s, and Fintushel and Stern show that this translates into a certain recursion relation involving the different terms in the desired formula for the invariants of a single blow-up. We will refer to the original paper for further details of the argument and move on to state the conclusion. In the single blow-up $X = X_1 \sharp \overline{\mathbf{CP}}^2$ the invariant $\langle E^d \rangle_X = \langle \mu(E)^d, M_X\rangle$ is a (finite) sum

$$\sum_i \frac{1}{d!} B_{d,i} \langle U^i, M^{4i}_{X_1}\rangle,$$

where $M^{4i}_{X_1}$ is a moduli space of dimension $4i$ over X_1 and the $B_{d,i}$ are universal numerical co-efficients. Let

$$B(t,x) = \sum_{d,i} B_{d,i} t^d x^i, \tag{8.1}$$

where x, t are initially formal variables. Fintushel and Stern's result is that if x, t are taken to be any complex numbers the sum in Equation 8.1 converges and the function $B(t,x)$ is

$$B(t,x) = e^{-t^2 x/3} \sigma_3(t), \tag{8.2}$$

where σ_3 is a Weierstrass sigma-function associated with the equation

$$\left(\frac{d\wp}{dt}\right)^2 = 4\wp^3 - g_2\wp - g_3, \tag{8.3}$$

with $g_2 = (16x^2 - 12)/3, g_3 = (64x^3 - 72)/27$!

We should recall here that our basic 4-dimensional class U differs by a factor of 2 from the more usual one, and this factor has to be brought in when comparing our formulae with others in the literature.

Let us try to understand the meaning of this remarkable result in terms of the original geometric data. We refer to [48] for information about these classical special functions, and we will use the notation of that text. First, the definition of the sigma-function. Given $g_2, g_3 \in \mathbf{C}$ such that the cubic

$$P(y) = 4y^3 - g_2y - g_3$$

has distinct roots in $\mathbf{C}$ we get an elliptic curve $z^2 = P(y)$. The expression $\frac{dy}{\sqrt{P(y)}}$ defines a holomorphic form ϕ on the curve, whose periods give a lattice $\Lambda \subset \mathbf{C}$. From this lattice we construct the familiar function $\wp(t)$ which satisfies Equation 8.3. Then we define $\zeta(t)$ by the conditions that $\zeta'(t) = -\wp(t)$ and

$$\zeta(t) = t^{-1} + 0 + at + \cdots$$

for small t. Next we define the sigma-function by

$$\frac{\sigma'}{\sigma} = \zeta$$

and

$$\sigma(t) = t + bt^2 + \cdots$$

for small t. This function $\sigma(t)$ is an odd function of t. The function $\sigma_3(t)$ that we need is an even function which is obtained as follows. Choose a basis $2\omega_1, 2\omega_2$ for the lattice Λ; then

$$\sigma_3(t) = Ce^{-\eta_3 t}\sigma(t - \omega_1 - \omega_2), \tag{8.4}$$

where C and η_3 are constants which can be fixed by the requirements that σ_3 is even and $\sigma_3(0) = 1$. Notice that, unlike σ, the function σ_3 is not completely determined by the original polynomial $F(y)$ since we need to make this choice of basis. More invariantly, the choice involved is a non-zero element of order 2 in the curve, or equivalently a root of P. So the function σ_3 is determined by a cubic polynomial P, with distinct roots, and a choice of a root.

We now bring in the connection with theta-functions. For any q in the unit disc we define

$$\theta_3(z;q) = \sum_{n=-\infty}^{\infty} q^{n^2} e^{2nz} = 1 + 2q\cos 2z + 2q^4 \cos 4z + 2q^9 \cos 6z + \cdots. \tag{8.5}$$

Similarly we put

$$\begin{aligned}\theta_1(z;q) &= 2\sum_{-\infty}^{\infty}(-1)^n q^{(n+1/2)^2} e^{(2n+1)z} \\ &= 2q^{1/4}\sin z - 2q^{9/4}\sin 3z + 2q^{25/4}\sin 5z - \cdots.\end{aligned}$$

Suppose we have chosen a basis for the lattice Λ as above. Then, according to (21.43) in [48], the sigma-function can be expressed as follows.

$$\sigma(z) = \frac{2\omega_1}{\pi\theta_1'} \exp\left(-\frac{\theta_1'''\nu^2}{6\theta_1'}\right)\theta_1(\nu; e^{\pi i\omega_2/\omega_1}). \tag{8.6}$$

Here $\nu = \pi z/2\omega_1$. The symbols θ_1', θ_1''' denote the derivatives of θ_1 with respect to z, evaluated at $z = 0$ and $q = e^{i\pi\omega_2/\omega_1}$. It is also shown on page 471 of [48] that

$$\frac{\theta_1'''}{\theta_1'} = 24\sum_{n=1}^{\infty}\frac{q^{2n}}{(1-q^{2n})^2} - 1. \tag{8.7}$$

Using the definition of σ_3, with this choice of basis for Λ, one finds that

$$\sigma_3(z) = \frac{1}{\theta_3(0;\omega_2/\omega_1)} \exp\left(-\frac{\nu^2\theta_1'''}{6\theta_1'}\right)\theta_3(\nu; e^{i\pi\omega_2/\omega_1}). \tag{8.8}$$

It is important to notice that this description depends on a choice of basis for Λ; different bases give different descriptions, via theta-functions, of the same sigma-function. In fact the assertion that the different formulae describe the same sigma-function is precisely the 'modular' property of the theta-function, in the q-variable.

We now turn to the case when

$$g_2 = \frac{16x^2 - 12}{3}, \quad g_3 = \frac{64x^3 - 72}{27}. \tag{8.9}$$

The reader will readily check that the roots of $F(y) = 4y^3 - 4g_2y - g_3$ are $-2x/3$ and $x/3 \pm \sqrt{x^2-1}$. (The first of these is the distinguished root used to fix the ambiguity in the definition of σ_3.) The roots are distinct so long as $x \neq \pm 1$. It will be convenient to think primarily of x as being real and large (although of course the formulae will extend to complex values by analytic continuation). The three roots are then all real and are ordered as

$$-\frac{2x}{3} < x_- < x_+$$

where

$$x_- = \frac{x}{3} - \sqrt{x^2 - 1}, \quad x_+ = \frac{x}{3} + \sqrt{x^2 - 1}.$$

We next make a choice of basis for Λ. The pre-image of the interval $[-2x/3, x_-]$ in the Riemann surface S is an embedded circle Γ_1. Likewise the pre-image of the interval $[x_-, x_+]$ is a circle Γ_2. With suitable orientations Γ_1, Γ_2 give a standard basis for the homology and the corresponding integrals are

$$\omega_1 = \int_{-2x/3}^{x_-} \frac{dy}{\sqrt{P(y)}}, \quad \omega_2 = \int_{x_-}^{x_+} \frac{dy}{\sqrt{P(y)}}. \tag{8.10}$$

Then ω_1 is real and positive, and we choose the sign of the square root so that ω_2 is a positive multiple of i: thus $\tau = \omega_2/\omega_1$ lies in the upper half-plane. By a linear change of variable, taking $-2x/3$ to 0 and x_+ to 1, we can write

$$\omega_1 = \frac{\pi}{2\sqrt{a}} F(\lambda), \quad \omega_2 = \frac{\pi}{2\sqrt{a}} G(\lambda), \tag{8.11}$$

where

$$F(\lambda) = \frac{2}{\pi} \int_0^\lambda \frac{dw}{\sqrt{w(1-w)(\lambda - w)}}, \quad G(\lambda) = \frac{2}{\pi} \int_\lambda^1 \frac{dw}{\sqrt{w(1-w)(\lambda - w)}}, \tag{8.12}$$

$$a = x + \sqrt{x^2 - 1}, \quad \lambda = \frac{x - \sqrt{x^2 - 1}}{x + \sqrt{x^2 - 1}}. \tag{8.13}$$

Our aim is to derive series representations for ω_1 and $q = e^{\pi i \omega_2/\omega_1}$ as $x \to \infty$. Clearly $\lambda \sim 1/4x^2$ as $x \to \infty$, so we want to expand in the small parameter λ. This is easy for the function $F(\lambda)$. Making a change of variable we write

$$F(\lambda) = \frac{2}{\pi} \int_0^1 \frac{dw}{\sqrt{w(1-w)(1 - \lambda w)}},$$

then put $w = \sin^2 \theta$ and expand

$$(1 - \lambda w)^{-1/2} = 1 + \frac{1}{2} \lambda w + \frac{1 \cdot 3}{2^2 2!} \lambda^2 w^2 + \cdots,$$

then integrate term by term using the fact that

$$\int_0^{\pi/2} \sin^{2k}\theta \, d\theta = \frac{\pi}{2}\frac{1\cdot 3\cdot\cdots\cdot(2k-1)}{2^k k!}.$$

This gives

$$F(\lambda) = (1 + a_1^2\lambda + a_2^2\lambda^2 + \cdots),$$

where

$$a_k = \frac{1\cdot 3\cdot\cdots\cdot(2k-1)}{k!2^k}.$$

Thus $F(\lambda)$ is the standard hypergeometric function $F(\frac{1}{2}, \frac{1}{2}; 1; \lambda)$ (in the notation of [48]). Using this we could write down a development of ω_1,

$$\omega_1 = \frac{\pi}{2\sqrt{2x}}\left(1 + \sum_{i\geq 1} b_i x^{-2i}\right)$$

where the co-efficients b_i are derived in a completely straightforward, but laborious, way from Equations 8.11 and 8.13. The first few terms of this series are

$$\omega_1 = \frac{\pi}{2\sqrt{2x}}\left(1 + \frac{3}{16x^2} + \frac{73}{2^{10}x^4} + \cdots\right).$$

The function $G(\lambda)$ is more difficult to handle because it has a logarithmic term as $\lambda \to 0$. If we allow λ to take complex values and to move around the origin the integral defining $G(\lambda)$ is changed after one circuit by subtracting $2F(\lambda)$, due to the different branches of the square root involved in the definition. It follows then that

$$G(\lambda) = -\frac{i}{\pi}\log\lambda \; F(\lambda) + H(\lambda),$$

where $H(\lambda)$ is analytic near 0. To find the series we can use the fact that $G(\lambda)$ satisfies the same hypergeometric equation as $F(\lambda)$. Now, from page 252 of [23], a second solution of this equation is $\log\lambda F(\lambda) + 2R(\lambda)$ where $R(\lambda)$ has the power series

$$a_1^2(2-1)\lambda + a_2^2\left(\left[2+\frac{2}{3}\right] - \left[1+\frac{1}{2}\right]\right)\lambda^2 + a_3^2\left(\left[2+\frac{2}{3}+\frac{2}{5}\right] - \left[1+\frac{1}{2}+\frac{1}{3}\right]\right)\lambda^3 + \cdots. \quad (8.14)$$

It follows then that

$$\pi G(\lambda) = -i\log\lambda \; F(\lambda) - 2iR(\lambda) + cF(\lambda)$$

for some constant c. In turn

$$q = e^{\pi i \omega_2/\omega_1} = \lambda e^{ic} e^{2R(\lambda)/F(\lambda)}.$$

Putting all this together, and except for the unknown co-efficient e^{ic}, we could write down a power series for q in inverse powers of x, as $x \to \infty$. To identify the co-efficient we can use the fact that the function $f(\lambda) = G(\lambda)/F(\lambda)$ is the inverse of the standard modular function J. That is f maps the upper half-plane to a hyperbolic triangle bordered by the imaginary axis, the line $\Re(\tau) = 1$ and the circle $|2\tau - 1| = 1$, and extends to a conformal equivalence from the universal cover of $\mathbf{C} \setminus \{0, 1\}$ and the upper half-plane. The function J has a product description

$$J(\tau) = 16q \prod_{n=1}^{\infty} \left(\frac{1 + q^{2n}}{1 + q^{2n-1}} \right)^8 . \tag{8.15}$$

Hence $\lambda = J(\tau) \sim 16q$ as $q \to 0$ so $q \sim \lambda/16$ as $\lambda \to 0$ and we see that $e^{ic} = 1/16$. (Of course another method of finding the power series for $q(\lambda)$ is to invert the power series for J resulting from Equation 8.15). So we conclude that

$$q = \frac{\lambda}{16} e^{2R(\lambda)/F(\lambda)}.$$

Putting everything together we can go back and substitute these power series for q and ω_1 into the formulae for the sigma-function and in turn for the blow-up function. The conclusion is that we can write

$$B(x,t) = p(x) e^{t^2 s(x)/8x} \theta_3(\sqrt{2x} r(x) t, h(x)/64x^2), \tag{8.16}$$

where p, h, r, s are power series in x^{-2}, with leading term 1. The first few terms are

$$p(x) = 1 - \frac{1}{32x^2} - \frac{3}{2^{10}x^4} + \cdots, \tag{8.17}$$

$$h(x) = 1 + \frac{1}{8x^2} + \frac{213}{2^{10}x^4} + \cdots, \tag{8.18}$$

$$r(x) = 1 - \frac{3}{16x^2} - \frac{161}{2^{10}x^4} + \cdots, \tag{8.19}$$

$$s(x) = 1 + \frac{5}{48x^2} + \cdots. \tag{8.20}$$

Now if we expand out the theta-function we find that the resulting sum contains certain terms which are precisely the 'naive' contributions we discussed in Chapter 7 – ignoring bubbling. One way of saying this is to replace the functions p, h, r by 1 (the leading term in the power series), and replace s by 0 (since it occurs inside the exponential with denominator $8x$) and define

$$\widehat{B}(x,t) = \theta_3(\sqrt{2x}t, 1/64x^2).$$

Now we expand $\widehat{B}$ as a Laurent series in x and polynomial in t

$$\widehat{B}(x,t) = \sum \frac{\widehat{B}_{nd}}{d!} x^n t^d,$$

and we find that the co-efficients $\hat{B}_{nd}$ give exactly the naive gluing formula from Subsection 7.3.4. Thus the co-efficients in the power series p, h, r, s determine the extra contributions which arise from the bubbling on the boundary of the moduli spaces, after the fashion of the work of Orszvath, Leness *et al.* Another way of expressing this is to introduce a parameter ϵ and define

$$B(x,t;\epsilon) = p(x/\epsilon)e^{t^2\epsilon^2 s(x/\epsilon)/8x}\theta_3(\sqrt{2x}r(x/\epsilon), h(x/\epsilon)/64x^2);$$

then we can expand $B(x,t;\epsilon)$ in powers of ϵ,

$$B(x,t;\epsilon) = \sum_{n=0}^{\infty} B_n(x,t)\epsilon^n.$$

Clearly $B(x,t) = \sum B_n(x,t)$ while the first term B_0 is just the 'naive' term $\widehat{B}$. It is reasonable to expect that the term B_n is the contribution from the part of the boundary of the moduli space arising from n 'bubbling points'.

The discussion applies equally well to multiple blow-ups. First recall that we can associate a theta-function to any definite integral form. Let L be a free abelian group of rank g; write $V = L \otimes \mathbf{R} \cong \mathbf{R}^{\mathbf{g}}$ and let ( ) be a positive definite integral form on L. The associated theta-function is a function of two variables, $q, \underline{t}$ where q lies in the unit disc in $\mathbf{C}$ and $\underline{t}$ is in V.

$$\Theta(\underline{t};q) = \sum_{\lambda \in L} q^{(\lambda\cdot\lambda)} e^{2i(\lambda,\underline{t})}.$$

Let $X = X_1 \sharp X_2$ where X_2 is a connected sum of g copies of $\overline{\mathbf{CP}}^2$ and $b^+(X_1) > 1$. We define a formal power series in $\underline{t} \in H_2(X_2)$ by

$$B(\underline{t}) = \langle e^{\underline{t}} \rangle_X,$$

by which me mean that we expand $e^{\underline{t}} = 1 + \underline{t} + \frac{\underline{t}^2}{2} + \cdots$ and evaluate the corresponding products on the relevant moduli spaces over X. Then it is an easy consequence of the single blow-up formula above that $B(\underline{t})$ is convergent and has a representation $B(\underline{t}) = B(\underline{t}, x)$ (where the variable x refers to the action of the 4-dimensional class over X_2 as before) with

$$B(\underline{t}, x) = p(x)^g e^{\underline{t}^2 s(x)/8x} \Theta(r(x)\sqrt{2x}\underline{t}, h(x)/64x^2).$$

Here Θ is the theta-function of minus the intersection form on $H_2(X_1)$, and p, s, r, h are the same functions of x as before. Again, the function $\widehat{B}(\underline{t}, x)$ obtained by replacing p, r, h by 1 and s by 0 gives the naive gluing formula.

There is another, at least conjectural, aspect of these formulae. Recall that our approach above has been to find expansions of the functions for large x. The formula above achieves this, but from the point of view of this formula it is not at all obvious that $B(x, \underline{t})$ is actually defined and holomorphic for all $x \in \mathbf{C}$; for example this is certainly not true for the naive approximation $\widehat{B}(x, t)$. What this means is that if we derive a Laurent expansion $B(\underline{t}, x) = \sum_{n=-\infty}^{\infty} b_n(\underline{t}) x^n$ from the formulae above then the co-efficients $b_n(\underline{t})$ for negative n are identically zero. This gives a system of identities satisfied by the intersection form, i.e. the standard diagonal form. To see this explicitly write $P_{a,l,m}$ for the polynomial function of $\underline{t}$

$$P_{a,l,m} = \frac{1}{(2a)!m!} \sum_{\lambda} (\underline{t}, \lambda)^a (\underline{t}^2)^m,$$

where the sum runs over vectors λ in the lattice with $(\lambda, \lambda) = l$. Then we can expand the theta-function to write

$$B(\underline{t}, x) = \sum A_{a,l,m,j} (-1)^a 2^{3(a-m-2l)} x^{a-m-2l-2j} P_{a,l,m}(\underline{t}),$$

where $A_{m,l,a,j}$ is the co-efficient of x^{-2j} in the expansion of

$$p(x)^g s(x)^m h(x)^l r(x)^{2a}.$$

The identities we obtain are the vanishing of the co-efficients of negative powers of x. For example the vanishing of b_{-1} gives, taking the part of degree 2 in the variable $\underline{t}$, the identity $2P_{0,0,1} = P_{1,1,0}$, i.e.

$$2(\underline{t}, \underline{t}) = \sum_{(\lambda,\lambda)=1} (\lambda, \underline{t})^2. \tag{8.21}$$

Taking the part of degree 6 in b_{-1} gives

$$-P_{3,2,0} + P_{2,2,1} + (2g - 32)P_{3,1,0} = 0.$$

The vanishing of the part of b_{-2} of degree 0 in $\underline{t}$ gives

$$2g = \sum_{(\lambda,\lambda)=1} 1. \tag{8.22}$$

The point is that these identities are true for the standard form, although less obviously so when the degree increases, but definitely not for all non-standard forms.

The first application of Yang–Mills instantons to 4-manifold theory was to show that the only definite intersection forms realised by the smooth 4-manifolds are the standard ones. The strategy was to take a moduli space M over such a manifold, truncate it to a compact manifold-with-boundary M_0 and then assert that

$$\langle \phi, \partial M_0 \rangle = 0 \tag{8.23}$$

for a suitable cohomology class ϕ defined by the familiar construction. Different moduli spaces and cohomology classes could be applied to produce essentially the same result: for example the argument of [11] (which fits into this framework if one uses the cobordism invariance of the Pontryagin numbers, and the fact that the first Pontryagin class of a moduli space is a multiple of our standard 4-dimensional generator), of [12] and of [19]. In principle if one takes any moduli space M one can identify components of ∂M_0 and write the identity Equation 8.23 as equalities between sums of different contributions from various pieces. (Of course there is no immediate motivation for this from 4-manifold theory, once one has already proved that all intersection forms are standard.) The difficulties in finding these individual contributions are much the same as in the direct attack on the blow-up problem: calculations of some terms, analogous to those of Orszvath and Leness, were made by Selby in his thesis [41]. In the light of the discussion above, it is natural to conjecture that the relevant identities are precisely those which arise from the vanishing of the negative terms $B_n(\underline{t})$. For example the arguments of [11] exploits the identity Equation 8.22 and that of [12] the identity Equation 8.21. One can extend the whole discussion of this Section to $U(2)$ bundles and one recovers similarly the argument of [19] as the first term in the corresponding expansion.

We now change back to the picture of a generalised connected sum $X = X_1 \sharp_Y X_2$ and take up again the discussion of Subsection 7.3.4. We will modify our notation slightly from that Chapter, to fit in the preceding discussion. Essentially we want to use the exponentials $e^{\underline{t}}$ rather than powers α^d. Thus, in this new notation, we saw in Chapter

7 that we can define for each class $\underline{t} \in H_2(X_2)$ a Floer chain $\psi(e^{\underline{t}}) \in CF_*(Y)$. More precisely we have a formal power series on $H_2(X_2)$ with values in the Floer chains, but we assume for simplicity that this is convergent. As we have explained in Chapter 7 we would like to define an object $F(\underline{t}, x)$ (which we think of as a function of two variables $\underline{t}, x$ or as a formal power series in $\underline{t}$ with values in power series in x), such that $\psi(e^{\underline{t}}) + F(\underline{t}, x)$ is a cycle in $\underline{\underline{CF}}_*$ and so defines an element $\underline{\underline{\psi}}(\underline{t})$ of $\underline{\underline{HF}}_*(Y)$. More precisely we would like the homology class to be independent of metric and there to be a gluing formula with respect to generalised connected sums. Now define

$$G(\underline{t}, x) = p(x)^g \exp(\underline{t}^2 s(x)/8x)\Theta_{X_2}(r(x)\sqrt{2x}\underline{t}; h(x)/64x^2) \qquad (8.24)$$

where p, r, h, s are the same functions as before, g is the second Betti number of X_2, and Θ_{X_2} is the theta-function of the intersection form on $H_2(X_2)$. Let

$$G(\underline{t}, x)_+$$

be the power series obtained by expanding G as a Laurent series in x and then discarding all the terms in negative powers of x. So $G(\underline{t}, x)_+$ is a power series in x and t. Our discussion of Fintushel and Stern's blow-up formula suggests that the solution to our problem from Chapter 7 might be to take

$$F(\underline{t}, x)_+ = G(\underline{t}, x)_+. \qquad (8.25)$$

This would fit in with the model case when $Y = S^3$, as we have explained above. (Of course Equation 8.25 is only a conjecture: it may be that the correct formula is a variant of this, given for example by modifying the functions p, h, r, s in some way.)

There is also an analogue of the discussion of the previous paragraph going back to Chapter 7. Let us assume for the moment that the picture suggested above is correct so we have a topological invariant $\underline{\underline{\psi}}(e^{\underline{t}}) \in \underline{\underline{HF}}_*(Y)$. Recall that we have an endomorphism u of $\underline{\underline{HF}}_*(Y)$ and a natural map R from $\underline{\underline{HF}}_*(Y)$ to $\mathbf{Q}[y]$. By definition the terms of degree greater than k in $R(u^k(f(\underline{t}))$ are just the terms of $x^k G(\underline{t}, x)$. So it is reasonable to expect that

$$R(u^k \underline{\underline{\psi}}(e^{\underline{t}})) = \left[x^k G(\underline{t}, x)\right]_+ . \qquad (8.26)$$

This would be a natural generalisation of Proposition 7.15. Let us assume that Equation 8.26 is correct. In contrast to the case of the standard form the function $G(\underline{t}, x)$ may now have non-trivial negative terms in its Laurent expansion, the precise form of the poles depending only on the intersection form of X_2. The Equation 8.26 yields a set of constraints on the ordinary Floer groups of Y, depending on the intersection form on X_2 of which Propositions 7.3 and 7.15 are prototypes. If this picture is correct, these constraints could be regarded as the proper generalisation of the theorem on the non-realisability of non-standard forms by closed 4-manifolds, to the case of 4-manifolds with boundary.

Bibliography

[1] J. F. Adams, *Lectures on Lie Groups* University of Chicago Press 1982

[2] M. F. Atiyah, Topological quantum field theories *Math. Publ. I.H.E.S.* **68** 175–186 1988

[3] M. F. Atiyah, New invariants of three and four dimensional manifolds In *The Mathematical Heritage of Hermann Weyl* Proc. Symp. Pure Math. 48 Amer. Math. Soc. 285–299 1988

[4] M. F. Atiyah, V. Patodi and I. M. Singer, Spectral asymmetry and Riemannian geometry *Math. Proc. Camb. Phil. Soc.* **362** 425–461 1978

[5] D. M. Austin and P. J. Braam, Morse–Bott theory and equivariant cohomology In *The Floer Memorial Volume*, Ed. Hofer *et al.* Birkhäuser 123–183 1995

[6] D. M. Austin and P. J. Braam, Equivariant Floer theory and gluing Donaldson polynomials *Topology* **35** 167–201 1996

[7] R. Bott, Morse theory indomitable *Math. Publ. I.H.E.S.* **68** 99–114 1988

[8] P. J. Braam and S. K. Donaldson, Floer's work on instanton homology, knots and surgery In *The Floer Memorial Volume*, Ed. Hofer *et al.* Birkhäuser 195–256 1995

[9] P. J. Braam and S. K. Donaldson, Fukaya–Floer homology and gluing formulae for polynomial invariants In *The Floer Memorial Volume*, Ed. Hofer *et al.* Birkhäuser 257–282 1995

[10] S. S. Chern, *Complex Manifolds without Potential Theory* Springer 1979

[11] S. K. Donaldson, An application of gauge theory to four-dimensional topology *Jour. Differential Geometry* **18** 279–315 1983

[12] S. K. Donaldson, Connections, cohomology and the intersection forms of four-manifolds *Jour. Differential Geometry* **24** 275–341 1986

[13] S. K. Donaldson, The orientation of Yang–Mills moduli spaces and four-manifold topology *Jour. Differential Geometry* **26** 397–428 1987

[14] S. K. Donaldson, Irrationality and the h-cobordism conjecture *Jour. Differential Geometry* **26** 141–168 1987

[15] S. K. Donaldson, Polynomial invariants for smooth four-manifolds *Topology* **29** 257–315 1990

[16] S. K. Donaldson, The Seiberg–Witten equations and 4-manifold topology *Bull. Amer. Math. Soc.* **33** 45–70 1996

[17] S. K. Donaldson and P. B. Kronheimer, *The Geometry of Four-Manifolds* Oxford University Press 1990

[18] J.-P. Eckmann and C. Wayne, Propagating fronts and the centre-manifold theorem *Commun. Math. Phys.* **136** 285–307 1991

[19] R. Fintushel and R. Stern, Definite four-manifolds *Jour. Differential Geometry* **28** 133–141 1988

[20] R. Fintushel and R. Stern, The blow-up formula for Donaldson invariants *Annals of Math.* **143** 529–546 1996

[21] A. Floer, Morse theory for Lagrangian intersections *Jour. Differential Geometry* **28** 513–547 1988

[22] A. Floer, An instanton invariant of 3-manifolds *Commun. Math. Phys.* **118** 215–240 1989

[23] A. R. Forsyth, *A Treatise on Differential Equations* Macmillan 1929

[24] D. Freed and K. Uhlenbeck, *Instantons and Four-Manifolds* MSRI Publications, Vol. 1, Springer 1984

[25] K. Froyshov, Equivariant aspects of Yang–Mills Floer theory *Topology* (To appear)

[26] K. Fukaya, Floer homology for connected sums of homology 3-spheres *Topology* **35** No. 1 89–136 1996

[27] R. Gompf and T. Mrowka, Irreducible four-manifolds need not be complex *Annals of Math.* **138** 61–111 1993

[28] D. Kotschick and J. W. Morgan, $SO(3)$-invariants for 4-manifolds with $b^+ = 1$, II *Jour. Differential Geometry* **39** 433–453 1994

[29] P. B. Kronheimer and T. S. Mrowka, Gauge theory for embedded surfaces, I *Topology* **32** 773–826 1993

[30] S. Lang, *Real Analysis* Addison-Wesley 1969

[31] T. Leness, Blow-up formulae for $SO(3)$ Donaldson polynomials *Math. Z.* **227** 1–26 1998

[32] W. Li, Floer homology for connected sums of homology 3-spheres *Jour. Differential Geometry* **40** 129–154 1994

[33] R. Lockhart and R. McOwen, Elliptic differential operators on non-compact manifolds *Annali della Scuola Normale Superiore di Pisa* 409–448 1985

[34] S. Lojasiewicz, *Ensembles semi-analytiques* I.H.E.S. notes 1965

[35] J. Morgan and T. Mrowka, A note on Donaldson's polynomial invariants *Int. Math. Research Notices* 223–230 1992

[36] J. Morgan, T. Mrowka and D. Ruberman, *The L^2-Moduli Space and a Vanishing Theorem for Donaldson Polynomial Invariants* International Press 1994

[37] V. Muñoz, Fukaya–Floer homology of $\Sigma \times S^1$ and applications *Jour. Differential Geometry* **53** 279–326 1999

[38] P. Orzvath, Some blow-up formulae for $SU(2)$ Donaldson invariants *Jour. Differential Geometry* **40** 411–447 1994

[39] A. Pressley and G. Segal, *Loop Groups* Oxford University Press 1986

[40] L. Schiff, *Quantum Mechanics* McGraw-Hill 1968

[41] M. Selby, D. Phil Thesis Oxford 1997

[42] L. Simon, Asymptotics for a class of nonlinear evolution equations with applications to geometric problems *Annals of Math.* **118** 525–571 1983

[43] E. M. Stein, *Singular Integral Operators and Differentiability Properties of Functions* Princeton University Press 1970

[44] C. H. Taubes, Self-dual Yang–Mills connections over non-self-dual four-manifolds *Jour. Differential Geometry* **17** 139–170 1982

[45] C. H. Taubes, Gauge theory on asymptotically periodic 4-manifolds *Jour. Differential Geometry* **25** 363–430 1986

[46] C. H. Taubes, Casson's invariant and gauge theory *Jour. Differential Geometry* **31** 547–599 1990

[47] C. H. Taubes, *L^2-Moduli Spaces on 4-Manifolds with Cylindrical Ends* International Press 1993

[48] E. T. Whittaker and G. N. Watson, *A Course of Modern Analysis* Cambridge University Press 1927

[49] E. Witten, Supersymmetry and Morse theory *Jour. Differential Geometry* **17** 661–692 1982

[50] E. Witten, Topological quantum field theory *Commun. Math. Phys.* **117** 353–386 1988

[51] E. Witten, Monopoles and 4-manifolds *Math. Res. Letters* **1** 769–796 1994

Index

3-torus, 29, 107

acyclic connection, 25
adapted bundle, 47, 51, 58, 130, 149, 160
admissible bundle, 147, 177
ASD connections, 9
Atiyah, M., 2

b^+, 10, 165, 169, 174
bubbling, 12, 114, 138, 196, 218, 226

centred instanton, 114
chain convergence, 116, 154
chambers, 163
Chern class, 10, 16
Chern–Simons theory, 16, 19, 23, 36, 202
Chern–Weil theory, 10, 17
cobordism, 2, 17
complex, 9, 151, 154, 160, 169, 183, 186, 188, 206, 215
conformal invariance, 9, 11, 14, 31
curvature, 8

elliptic operator, 9, 23, 33, 40, 44, 48
eta-invariant, 66
expected dimension, 12, 109, 161

Fintushel, R., 6, 219
Floer, A., 1, 30, 136
Fredholm index, 9, 13, 51, 59, 61
Froyshov, K., 5, 149, 219
Fukaya, K., 5, 201
Furuta, M., 6

gauge group, 7, 84, 86
gluing parameter, 101, 163, 173, 197, 212
gluing techniques, 54, 91, 100, 108, 112, 128, 157, 163, 173, 202, 219, 226
gradient, 32, 34, 142

h-cobordism, 124, 151
Hessian, 23, 151
Hodge theory, 8, 24
holonomy, 7, 135, 179, 202
homology, 5, 133, 151
homology 3-sphere, 113, 159

instanton, 1, 4, 9, 11, 20, 30, 39
intersection form, 10, 175, 201, 228, 230
invariant, 161
irreducible connections, 12, 16, 20, 24

Kotschick, D., 6, 163
Kronheimer, P., 4
Kuranishi map, 28, 109, 142

L^2-compatible, 28, 107
Laplace operator, 23, 26
Leness, T., 219, 226

moduli space, 12, 13, 82, 87, 100
Morgan, J., 5, 161, 163
Morse function, 157
Morse theory, 1, 27, 29, 151, 177, 194
Morse–Bott function, 107, 108, 194, 208
Mrowka, T., 4, 5, 161
Muñoz, V., 5

non-degenerate connection, 25

orientation, 13, 130
Orszvath, P., 219, 226

quantum field theory, 1, 38

reduced moduli space, 114, 180

reducible connection, 7, 154, 161, 168, 174, 203
regular point, 86, 111, 145
representation of π_1, 10, 20

Segal, G., 2, 157
Seiberg–Witten theory, 4
Simon, L., 106
slice, 8, 11, 86
Sobolev inequality, 7, 45, 72, 75, 94, 95, 98
spectral flow, 66
Stern, R., 6, 219

Taubes, C., 5, 92
temporal gauge, 15, 81
theta-function, 222, 226
topological quantum field theory, 2, 151, 159, 168
transversality, 13, 135, 142
tubular end, 13, 40

Uhlenbeck, K., 11, 14, 87

virtual dimension, 12, 99

weak convergence, 12, 114
weighted spaces, 58, 88, 100, 109, 205
Witten, E., 1, 153

Yang–Mills, 7, 14, 35

Printed in the United States
By Bookmasters